Radio Astronomy for the Amateur

DEDICATION

This book is dedicated to my wife, Judy, and all of my neighbors whose patience and understanding allowed me to turn my suburban backyard into an antenna farm.

No. 714
$8.95

Radio Astronomy for the Amateur

by Dave Heiserman

TAB BOOKS
Blue Ridge Summit, Pa. 17214

FIRST EDITION

FIRST PRINTING—APRIL 1975
SECOND PRINTING—JANUARY 1977

Printed in the United States
of America

Hardbound Edition: International Standard Book No. 0-8306-5714-2

Paperbound Edition: International Standard Book No. 0-8306-4714-7

Library of Congress Card Number: 74-33624

PREFACE

Radio amateurs and electronics experimenters are forever looking for new ideas to whet their appetites for building and using novel electronic circuits. Anyone with a technician-level background in modern electronics and a curiosity about the universe around him will find the projects described here to be most challenging and satisfying to his imagination.

The presentations assume the reader has mastered the basic skills of circuit construction. On the other hand, it is assumed that he has little or no formal knowledge of astronomy.

Chapters 2, 3, and 4 are devoted entirely to the elements of astronomy that have a direct bearing upon the needs of amateur radio astronomers. The reader is encouraged to seek out further basic knowledge of astronomy from books dealing exclusively with that subject.

Chapters 5 through 11 show an amateur experimenter what he can do with radio astronomy and describe how he can go about doing it.

The author is indebted to a number of individuals who helped with the preparation of this book. A special word of appreciation must go to the entire staff of the Ohio State University radio telescope installation at Delaware, Ohio. I am also grateful to my friend and neighbor, William Howells, who helped with much of the labor involved in constructing some of the more elaborate antenna systems.

Dave Heiserman

CONTENTS

A Perspective on Radio Astronomy

1

It is highly likely that astronomy is man's first real science. Only bits and pieces of the first accounts of astronomical studies are available today, but there can be little doubt the techniques of this prehistoric science set the pace for the development of other ancient sciences, including mathematics and physics.

Radio, on the other hand, is a relatively new science, and its beginnings in the mid-1800s are thoroughly documented in scientific papers. Whereas astronomy developed into a mature science over a span of thousands of years, radio grew into an important science and a vital technology in less than one century.

Radio astronomy is a wholly modern science that has well defined beginnings in the years between 1929 and 1931. Instead of requiring thousands of years or even several decades to mature, radio astronomy blossomed into a workable science within a short 15 years.

THE BEGINNINGS OF RADIO ASTRONOMY

As radio technology developed, it became clear that long-distance transmissions were overly sensitive to sources of unwanted electrical noise. Two kinds of electrical interference were easy to study: lightning and man-made electrical noise. Radio scientists and engineers soon tracked down a source of natural radio interference caused by the

Sun's rays striking charged particles in the Earth's upper atmosphere.

Engineers at Bell Telephone Laboratories wanted to set up reliable transoceanic radiotelephone links to supplement an underwater cable system, and in the late 1920s launched a research program to study the ways the Sun and other sources of interference influence long-distance radio transmissions.

Bell Labs put one of its top radio engineers, Karl G. Jansky, in charge of the radio interference project. The basic idea of the project was to intercept, pinpoint, and identify sources of radio noise. Unlike most of the radio interference experiments conducted up to that time, Jansky and Bell Labs decided to leave transmitters out of the scheme, and merely *listen* with an array of antennas, receivers, and recording devices.

Working on a plot of land in New Jersey, Jansky built up several large omnidirectional antenna arrays and easily identified interference from distant thunderstorms and man-made electrical noise from nearby electrical power installations. He also studied what we know as the *E-layer* effects in the upper atmosphere; and he was able to pinpoint and track radio noise from the Sun as it moved across the sky.

Studying the frequency range around 20.6 MHz, Jansky accidentally came across another source of powerful radio static in the sky. His antenna in this instance was the wooden merry-go-round pictured in Fig. 1-1. This elaborate arrangement supported a set of 21-foot dipoles and reflectors, and rode around in a circle on four old Ford automobile wheels.

What Jansky had found was a powerful point of radio static that seemed at first to lead the Sun across the sky. This close relationship to the motion of the Sun at first led Jansky to believe the source was located in the Earth's upper atmosphere and that it was somehow linked to the E-layer effects of the Sun.

Jansky's preliminary theory about the origin of this newly found source of static soon became less certain and eventually fell completely apart. The noise source did, indeed, lead the Sun across the sky for several months; but as time passed, the point dropped behind the Sun, and gradually fell so far behind that it appeared in the sky during nighttime hours.

Jansky had only a layman's interest and knowledge of astronomy at the time, so perhaps it is understandable that he did not consider the possibility that the noise could be coming from a point far beyond the Earth's atmosphere and, in fact, outside the solar system. While discussing the puzzling "atmospheric effect" with an acquaintance who did know something about astronomy, Jansky learned he could try correlating the position of the strange noise source with the motion of the stars rather than the movement of the Sun.

After teaching himself some rudimentary celestial mechanics, Jansky found an exact correlation between the motion of the stars and his new radio source. The static was actually coming from a portion of the sky in the direction of *Sagittarius*—a constellation that stands between the Earth and the very center of our galaxy.

With the support of knowledgeable astronomers, Jansky established beyond any reasonable doubt that he had been tracking radio emissions from outer space. On April 27, 1933,

Fig. 1-1. Historic Bell Labs photo of the late Karl G. Jansky and the world's first radio telescope antenna.

Jansky formally announced his findings and gave birth to the modern science of radio astronomy. Newspapers and magazines had a heyday with the story. Some described the signals as transmissions from intelligent life in space, others treated it as the beginning of a new and legitimate science, and still others suggested the whole thing was a hoax.

The history of science is noted for the way fate intertwines fortunate and unfortunate circumstances. In Jansky's case, the discovery of radio emissions from space was a fortunate accident. The unfortunate part is that Jansky never lived to see his discovery turned into a mature science. Radio noise from the center of the galaxy posed no threat to long-distance radio communications, and Bell Labs displayed a rather uncharacteristic fit of shortsightedness—the company withdrew all program support and assigned Jansky to a different kind of project. Radio astronomy thus entered a period of stagnation that lasted until about 1940.

Grote Reber, a radio engineer by profession and an amateur astronomer and radio "ham" by inclination, became fascinated with Karl Jansky's discovery. Around 1939, Reber used his own money and ingenuity to build a large steerable parabolic antenna in his backyard in Wheaton, Illinois. The parabolic antenna promised to have more gain and directivity than Jansky's dipole array; and to keep the whole system scaled down to backyard proportions, Reber decided to boost the operating frequency to the newly opened VHF range.

The emissions from the galactic center were not as strong at 160 MHz as they were in the 20.6 MHz range Jansky used, but Reber's high-gain antenna and VHF receiver made up the difference.

Reber soon found that the galactic center was not the only source of extraterrestrial radio noise. He found a secondary maximum in the direction of the constellation *Cygnus*; and perhaps more importantly, he found a moderate and evenly distributed region of noise that followed the entire galactic plane, the *Milky Way*.

In 1940, Reber published a paper announcing the fact that the Milky Way was a source of radio noise; and by 1942 he was able to produce the first radio map of the sky.

The events of a world at war dulled the impact of Reber's work; but the development of VHF and UHF military communications equipment eventually had a powerful effect on later progress in radio astronomy.

Near the end of the war, Edward M. Purcell at Harvard University devised a technique for monitoring the vibrating magnetic fields generated by the atomic nucleus. His discovery later led to his winning the Nobel Prize; but more importantly as far as radio astronomy is concerned, Purcell suggested the stars might produce immense doses of the same kind of energy. More specifically, the atoms of hydrogen that fill the universe should emit powerful radio energy at a wavelength of precisely 21 centimeters.

In March 1951, Dr. Purcell and another Harvard professor, Harold Ewen, completed the first 21 cm radio telescope on the roof of their laboratory. The signals Purcell had predicted nearly six years before were there, and radio astronomy got the boost it needed to pull it out of its *dark ages*.

Fig. 1-2. This photo from the National Radio Astronomy University shows America's largest steerable parabolic antenna. The dish, located at Green Bank, West Virginia, has a surface area of 78,000 square feet; its resolution at 1420 MHz is about 10 minutes of arc.

Fig. 1-3. The world's most comprehensive radio astronomy mapping projects have been carried out with this antenna system at Ohio State University. Incoming signals are reflected from the movable plane in the upper right corner of this photo to the fixed plane in the lower left corner. The small building at one end of the "football field" contains the horn-type receiving antennas.

Things happened quickly after that. Aside from Jansky's original announcements, no other feat of radio astronomy caught the popular imagination as much as the construction of the giant 250 ft dish antenna at Jodrell Bank in England. The persuasive powers and enthusiasm of a single man, Sir Bernard Lovell, were largely responsible for bringing about the fantastic and continuing success of the Jodrell Bank project. The world's supergiants of science, the United States and the Soviet Union, have not had the benefit of such a powerful personality; and England stands as the leader in earthbound radio astronomy today.

That is not to say the the U.S. hasn't been involved in radio astronomy. The Green Bank, West Virginia, dish shown in Fig. 1-2 is a fairly impressive piece of workmanship. And some of the most ambitious mapping projects ever devised were undertaken and completed using the Ohio State University facility pictured in Fig. 1-3.

Introduction to General Astronomy

Radio astronomy brings together the ancient science of astronomy and the relatively new arts of electrical and mechanical engineering. Few professional radio astronomers are equally competent in both astronomy and engineering, so a radio astronomy group is normally made up of a number of engineers with some astronomy background and perhaps one or two astronomers working in an advisory capacity.

An amateur radio astronomer wanting to develop his own equipment must have a good working knowledge of the theory and practice of *communications* electronics. That is most important. But technical expertise is of no use at all without some knowledge of astronomy as well.

This chapter and the one that follows deal with the basic elements of conventional astronomy and the techniques commonly practiced by amateur astronomers. These discussions are not exhaustive by any means—the reader can consult a variety of popular astronomy books for further details. The objective here is to present some of the theories and techniques of optical astronomy that have direct relevance to amateur radio astronomy.

CONVENTIONAL VIEWS OF THE SKY

Philosophers during the Middle Ages believed the Earth rested at the center of the universe and that all the celestial bodies rotated about it. This rather presumptuous point of

Fig. 2-1. Celestial globe shows the earth surrounded by an imaginary celestial sphere containing all the heavenly bodies. (Photo courtesy of Edmund Scientific Co.)

view gradually gave way to the Copernican view that places the Sun at the center of the universe. Today we know the Copernican view isn't quite right, either. The Sun is indeed at the center of the solar system, but it is certainly not at the center of the entire universe. No one really knows where the center of the universe might be; and in fact there might be some doubt about whether the notion of a "center of the universe" has any real meaning at all. Perhaps future discoveries in radio astronomy will shed some light on the geometric shape of the universe, and thereby answer some perplexing cosmological questions.

When working practical problems in astronomy, however, astronomers usually adopt the ancient point of view. With so many celestial bodies rotating about so many different points in space, it is convenient to select one of the points as a stationary frame of reference, and assume all other motions are relative to it. For practical purposes, then, the Earth is often considered a stationary platform in space that has the Sun, Moon, planets, and all the stars moving about it.

The ancient philosophers also believed the heavenly bodies were all about the same distance from Earth. Today we know this is far from true. But for practical purposes we can think of all the stars as being situated on the inner surface of a great *celestial sphere*.

The Celestial Sphere

The celestial sphere is the inner surface of a great globe that has all the stars and planets painted on it. You can get the idea by examining Fig. 2-1, which shows the Earth surrounded by a great spherical surface of heavenly bodies. As the sphere revolves, the stars seem to move through circular paths around the Earth.

Figure 2-2 is a time-exposure photo of Polaris (the *North Star*) and some nearby stars. Such star-trail photos closely

Fig. 2-2. Time-exposure "star trail" photo of the night sky shows Polaris at the center of a set of arcs representing the apparent circular motion of the stars. (Lick Observatory photo.)

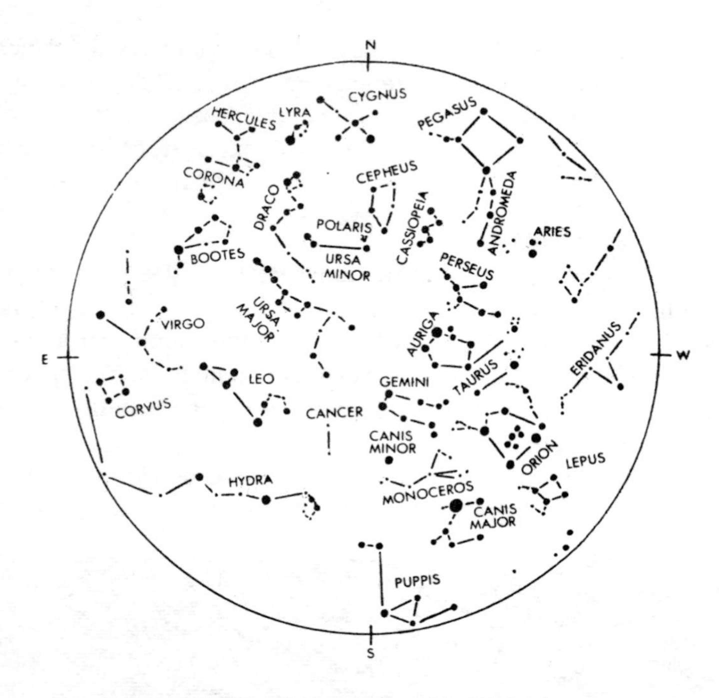

Fig. 2-3. Star Map I—See Table 2-1 for key.

resemble a view of the celestial sphere as seen from the inside. There is nothing in the picture, for example, that indicates the stars are all different distances from Earth. Further, there is no way to tell whether the trails are due to the Earth's rotation (which they really are) or caused by the rotation of the stars around the Earth.

Although celestial globes and star-trail photos give the user a very poor cosmological view of the universe, they serve as handy models for studying the motions of the stars and their angular positions relative to the Earth. Celestial globes are far too cumbersome and difficult to use when standing outside trying to locate certain stars and constellations in the night sky, however; and a simple star map can replace an elaborate celestial globe for such everyday applications.

Simple Star Maps

Star maps, like a celestial globe, do not show the actual sizes and distances of the stars, but they do show the star patterns in a way that makes it easy to find them on a clear, dark night.

Figures 2-3 through 2-8 show star maps as seen from most of the northern hemisphere. Such star maps are especially helpful to people learning their way around the sky. The maps do not have the precise and detailed information required for close astronomical work, but they show most of the basic 88 star patterns, or *constellations*, and their relative positions in the sky at different times of the year.

Star Map Distortions. Since star maps are two-dimensional representations of patterns that appear on the surface of a three-dimensional globe, they always reflect the same kind of distortion one finds on geographical maps. Constellations drawn near the center of circular star maps, for instance, are always more accurate than those appearing near the edges. Just as the continent of Antarctica appears immense compared to the United States on a typical geographical map, constellations situated near the edges of a

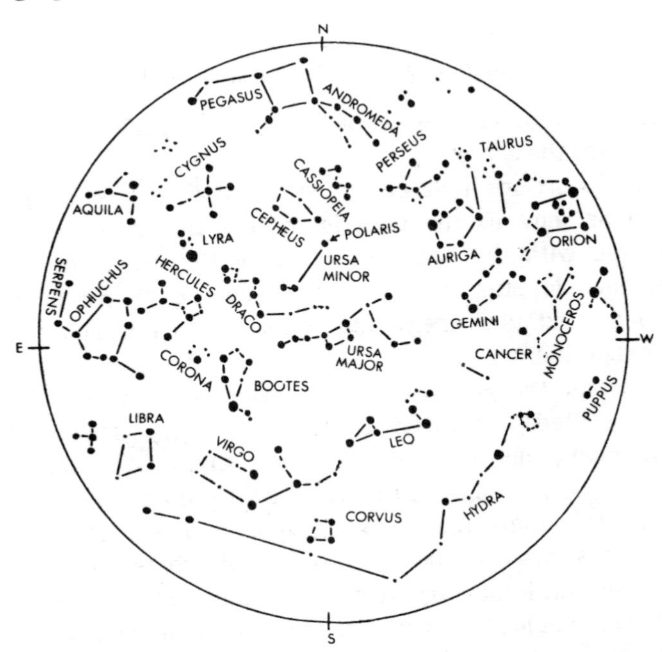

Fig. 2-4. Star Map II—See Table 2-1 for key.

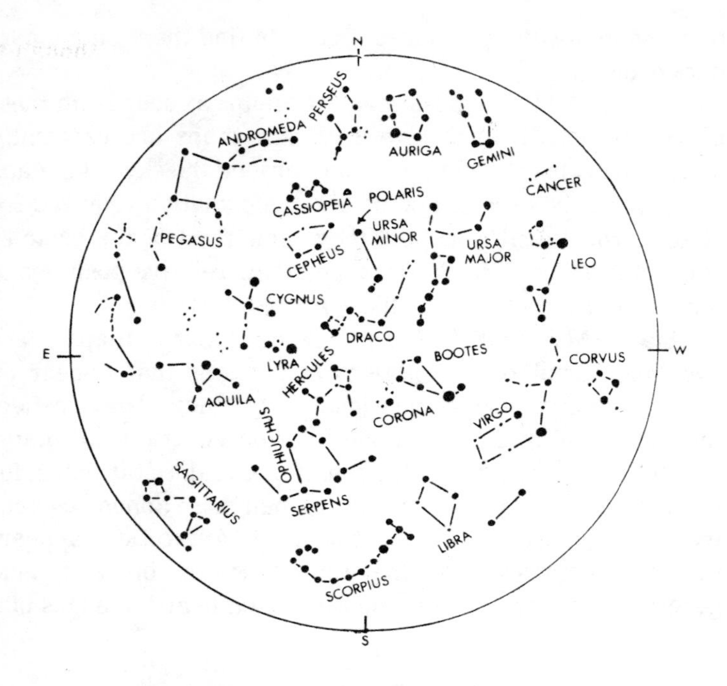

Fig. 2-5. Star Map III—See Table 2-1 for key.

circular star map are always blown out of proportion relative to those near the center.

The distortions inherent in star maps are of no real concern as long as the user is merely trying to become acquainted with the general picture of the nighttime sky. Knowledgeable star viewers are of course aware of the distortions and take them into account when working out practical problems.

Note that the geographical directions of east and west seem reversed compared to the way they appear on geographical maps. This reversal of east and west sometimes worries a beginner until he realizes the difference between the way a star map and a geographical map are used. Geographical maps of the Earth or its continents are always drawn from an imaginary viewing position above the Earth. The observer is looking downward at the Earth from the sky, making the eastward direction always appear 90° clockwise from north. The viewing position is just the other way around

the year. The Earth's daily rotation carries all members of the solar system and the stars across the sky from east to west. Only artificial satellites violate this general rule.

What is perhaps less obvious is the fact that objects in the solar system seem to move across the sky somewhat more slowly than the stars do. And generally speaking, the farther an object is from Earth, the more its motion matches that of the distant stars.

Since the positions of the stars in the sky change from hour to hour and from one day to the next, any one star map is accurate only for certain times and dates. And since the positions of the Sun, Moon, and planets change compared to the stars in a predictable but often complicated manner, simple star maps do not include members of the solar system.

Each of the six simple star maps in this chapter is reasonably accurate for six different times and dates. A time-and-date key for these maps appears in Table 2-1.

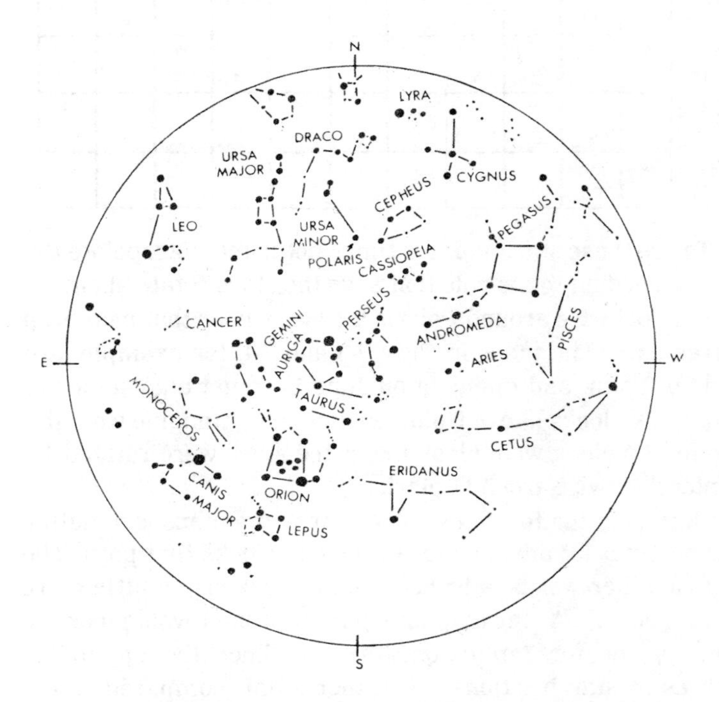

Fig. 2-8. Star Map VI—See Table 2-1 for key.

Table 2-1. Index to Star Maps I through VI

DATE	LOCAL STANDARD TIME 2000	2100	2200	2300	0000	0100	0200	0300	0400	0500	0600
JAN 1			VI				I				II
15		VI				I				II	
FEB 1	VI				I				II		
15				I				II			
MAR 1			I				II				III
15		I				II				III	
APR 1					II				III		
15				II				III			
MAY 1			II				III				IV
15		II				III				IV	
JUN 1	II				III				IV		
15				III				IV			
JUL 1							IV				V
15						IV				V	
AUG 1	III				IV						
15				IV				V			
SEP 1			IV				V				VI
15		IV				V				VI	
OCT 1	IV				V				VI		
15				V				VI			
NOV 1			V				VI				I
15		V				VI				I	
DEC 1	V				VI				I		
15				VI				I			

To use these star maps for times not listed, interpolate the times according to the rule that says the stars rotate about 15° counterclockwise around Polaris for every hour that passes on a given date. The maps in Figs. 2-3 and 2-4, for example, are good for 8 p.m. and midnight on June 1. To get a good idea of what the sky looks like at 9 p.m. on that date, imagine what the map for 8 o'clock would look like if the stars were rotated 15° counterclockwise around Polaris.

Interpolating for *dates* not shown on the maps is a matter of applying a 15° advance for every two weeks that pass. The stars, in other words, advance counterclockwise a little more than 1° per day. A star map for 8 p.m. on June 1 would look the same as one for 7 p.m. on June 15. Since the day-to-day changes in star positions are rather small compared to the hourly changes, casual star viewers normally use the map for the nearest date listed, and apply only the hourly correction.

Simple Star Maps and Geographical Latitude. The positions of the stars in the sky change with the observer's latitude as well as the time and date. A change in latitude, however, shifts the stars north or south rather than from east to west. The star maps shown in this chapter are intended for observers between 30° and 45°N latitude. Corrections for local latitude take advantage of the fact that the angle between Polaris and the northern horizon is equal to the observer's geographical latitude.

If, for instance, an observer is at 35°N latitude, Polaris will always appear 35° over the northern horizon. He thus knows his view of the sky matches that shown on a simple star map when the map is adjusted so that Polaris appears 35° over his northern horizon.

THE SOLAR SYSTEM

The present-day picture of the solar system places the Sun at the center and shows nine major planets orbiting about it. Five of the planets—Mercury, Venus, Mars, Jupiter, and Saturn—are readily visible at one time or another without using any kind of optical aid. Uranus and Neptune can be found with the help of a relatively good amateur optical telescope, but Pluto is quite elusive under any circumstances.

Standard information about the solar system can be found in any good astronomy book and in references such as encyclopedias and almanacs. This kind of information can be quite valuable to an amateur radio astronomer; but since it is so readily available from other sources, there is little point in dwelling on it here.

A beginning amateur radio astronomer might be more interested in answers to two important questions: What are the most interesting radio objects in the solar system, and how do I go about locating them and determining their positions in the past and in the future?

Radio Objects in the Solar System

The Sun and Jupiter are by far the most active radio objects in the solar system. The Sun, in fact, is normally listed as the most powerful radio source ever observed from Earth. Jupiter is ranked second, but the giant planet's radio output is known to exceed that of the Sun on occasions.

The remaining planets and natural objects in the solar system emit some radio energy, but the levels are so low that direct radio studies are virtually impossible with amateur equipment and procedures. It is conceivable, however, that some of these "silent bodies" could exert secondary effects upon radio emissions from the Sun and Jupiter. The position of the Jovian moon, *Io*, does indeed influence radio emissions from its host planet, while a solar eclipse and a transit of Mercury are likely to produce some effects upon solar radiation.

Chapter 3 outlines a number of other kinds of solar system studies that have occupied the interest of amateur optical astronomers for a number of years. By becoming acquainted with these optical projects (and perhaps trying a few first-hand), a radio amateur might well modify them for radio astronomy work.

An amateur radio astronomer will always find the Sun and Jupiter the most intense and reliable radio sources in the sky. Pointing the antenna toward the Sun is a good first test of a system's sensitivity and general operating condition. Jupiter can be a good target for a first test, too; but many beginners have trouble finding it in the sky.

Locating the Planets

Although Jupiter is the second brightest radio source in the sky, it normally takes fourth place in terms of optical brightness. Jupiter can be a rather inconspicuous object among the stars; and unless a beginning radio astronomer knows how to find Jupiter optically, he is bound to have some difficulty studying its radio characteristics.

The simplest and most straightforward way to locate a planet such as Jupiter is by consulting a monthly astronomical map that shows the positions of all the planets against a background of the major stars.

One good source of monthly maps is *Sky and Telescope* magazine. A small map in every issue shows the

Moon, Sun, and all the major planets on an abbreviated celestial map. Do not confuse the planetary map with the monthly centerfold that shows the positions of all visible stars and other major objects outside the solar system.

The planet maps in *Sky and Telescope* show the planets as they appear against the stars, and not as they would appear with respect to the observer's local horizon. The maps, in other words, do not directly show where a planet can be found from any one particular observation point on Earth. The user must first note the stars on the map that are closest to the planet he wants to find. He then has to determine the positions of those stars for his viewing point by consulting the star maps in this chapter. Once he locates the background stars in the sky, he can look for the planet in that general area.

A helpful hint on distinguishing planets from stars: planets well above the horizon do not twinkle as much as stars do.

An alternate technique for locating planets is by noting their times of rising and setting as listed in standard almanacs. If the planet happens to rise at 4 p.m. and set at 4 a.m. the following day, it follows that it will be in the sky for a total of 12 hours. If the desired time of observation is midnight, it figures that the planet will have been in the sky 8 of its full 12 hours. The planet will thus be $^{8}/_{12}$ or 75% of the way from the eastern to the western horizon.

Knowing a planet will be three-fourths of the way across the sky at a certain time, however, is not altogether helpful—there is a lot of sky three-fourths of the way to the western horizon. To determine the position of the planet with a bit more accuracy, the observer has to know the approximate position of the *ecliptic* at the time.

The ecliptic is a great circle on the celestial sphere that varies between 27° and 63° above the southern horizon from most points in the continental United States. The ecliptic actually defines the plane of Earth's orbit about the Sun; and since most of the other planets share pretty much the same plane, it follows they will always appear rather close to the ecliptic.

It is thus possible to reckon the approximate position of a planet by first figuring its east-to-west displacement from the listed times of rising and setting, and then looking for it near the ecliptic.

The two planet-finding techniques described so far are rather informal. In practice, these informal techniques are adequate for most amateur optical and radio astronomy work—especially when the target is one of the brighter planets.

On a clear night, Jupiter shines brighter than any of its background stars, so an observer can readily pick it out once he knows about where to look. The informal procedures for locating Jupiter are also good for daylight radio observations because radio telescope antennas for Jovian studies normally cover a 30–60° segment of the sky. If the observer's estimates of Jupiter's daylight position are anywhere close to being accurate, the system is bound to pick up some of its signals.

Experimenters desiring the highest possible degree of positional accuracy should consult *The American Ephemeris and Nautical Almanac* for the celestial coordinates of Jupiter. The *Ephemeris* is available at most public libraries, but the user must also be familiar with the kind of mathematical work described in Chapter 4 of this book.

OUR GALAXY

While Earth is a member of the solar system, the solar system is, in turn, a member of our galaxy; and our galaxy is only one of millions of similar "island universes" that are scattered throughout the entire known universe.

Our galaxy contains about 100 billion stars. Only about 5000 stars are distinguishable as individual points of light with the naked eye, and a good 8-inch optical telescope can resolve about 5 million of them. A good share of the stars in our galaxy are so distant that they are, in themselves, invisible. The overall effect of billions of such stars, however, creates the impression of a blue-white haze. The best-known "haze" of stars is the Milky Way.

A great deal of evidence accumulated over the past 25 years suggests our galaxy has a flat, spiral structure similar to the famous spiral galaxy in Andromeda (Fig. 2-9). Viewing our galaxy edge-on, it most likely looks like a pancake with a

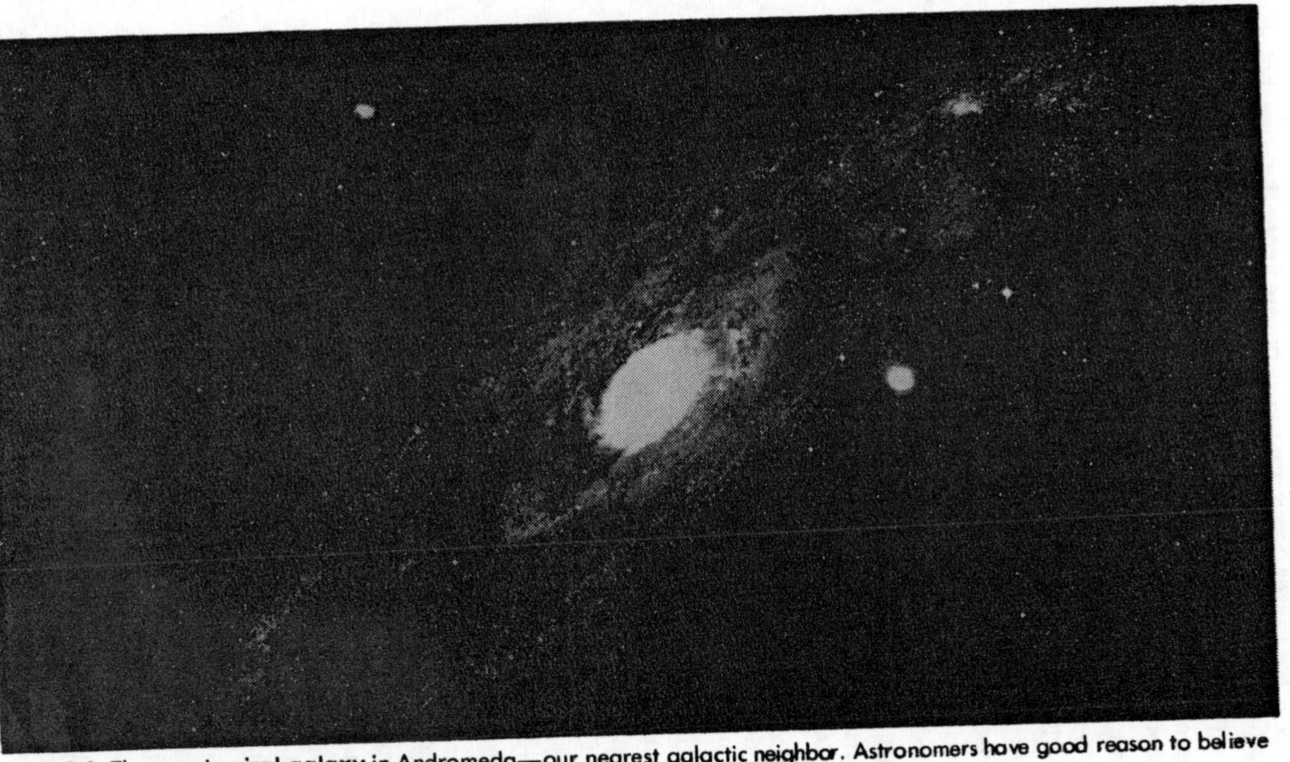

Fig. 2-9. The **great spiral galaxy** in Andromeda—our nearest galactic neighbor. Astronomers have good reason to believe our own galaxy has much the same appearance. (Lick Observatory photo.)

bulge in the center. Stars are well separated near the edge, but become increasingly clumped together near the center.

The tremendous population of suns near the galactic center produce gravitational and thermal havoc that makes our Sun's output seem puny by comparison. It is little wonder, then, that the galactic center should be one of the most powerful sources of radio energy observable from the Earth.

The solar system is located near the edge of our galaxy, and probably rests within one of the giant spiral arms. An observer looking at the Milky Way from Earth is actually peering into the plane of the galaxy. The center of the galaxy stands 32,000 light-years away in the direction of the constellation of Sagittarius. The outermost edge of the galaxy is in the Milky Way toward Taurus.

A simple radio telescope system is quite capable of detecting a general galactic background noise whenever it is pointed toward any part of the Milky Way. The galactic activity is most intense at the center, however. Karl Jansky carried out his pioneering work by tracking noise from the galactic center—a source now commonly known as *Sagittarius A*.

One of the real payoffs of radio astronomy has been the information gained about the galactic center. The galactic center is not visible optically, because of an immense cloud of gas that stands between it and the solar system. The gas cloud is transparent to radio-frequency radiation, however, and much of what we now know about the structure of our own galaxy comes from radio studies of the "invisible" galactic center.

The Milky Way and Sagittarius A (the galactic center) are technically known as *extended radio sources*. Such sources are made up of a huge number of discrete objects which, by themselves, would pass completely unnoticed. To a radio telescope system, extended sources appear to cover a relatively large segment of the sky.

By way of contrast, *discrete radio sources* cover only a very small amount of sky, and often can be associated with an optical object. The Sun and Jupiter are examples of discrete radio sources within the solar system, and there are two powerful discrete radio sources outside the solar system, but within our galaxy. To optical astronomers, these two discrete sources are known as *Tycho's Star* and the *Crab Nebula*. Radio astronomers know them as *Cassiopeia A* and *Taurus A*.

Cassiopeia A and Taurus A are very likely remnants of a supernova—a star that literally destroyed itself in a burst of energy that emitted immeasurable amounts of radiation across a broad frequency spectrum that included light, X-rays, and radio waves. Chinese astronomers recorded descriptions of the supernova in Taurus in 1054 AD. The famous Danish astronomer, Tycho Brahe, takes credit for studying the birth of Cassiopeia A in 1572.

Both of these supernovas flared up in brightness for several months before beginning to settle down. First-hand reports describe the objects as being brighter than Venus. The debris left over from the explosions is now visible as rather faint nebulosities through telescopes of 6 inches in diameter or more. They remain as powerful radio sources, however.

Cassiopeia A, in fact, is the third most powerful radio source observable from Earth, but actually generates nearly a million times as much radio energy as the Sun and Jupiter put together. Twelve-thousand light-years tend to lessen the intensity of any radio signal.

The signals from Taurus A (Crab Nebula) are not particularly strong, and an investigator using very simple equipment might have some trouble locating it. The galactic center, Sagittarius A, is slightly more powerful than Taurus A; so if the equipment can pick up Sagittarius with a reasonable degree of reliability, the experimenter should be able to locate Taurus when conditions are right.

It so happens the Sun passes very close to Taurus A during the middle part of June each year. Some amateur radio astronomers have spent a great deal of time noting how the Sun's radio corona masks signals from Taurus when the "eclipse" occurs.

EXTRAGALACTIC SOURCES

One of the most intriguing features of radio astronomy is the fact that it extends the observable universe far beyond the limits of optical astronomy. The four brightest optical objects—the Sun, Moon, Venus and Jupiter—are all within the solar system. Jupiter is the most distant of these bright objects, averaging 500 million miles from Earth.

The four "brightest" discrete radio objects are the Sun, Jupiter, Cassiopeia A, and *Cygnus A*. Cassiopeia A, the third brightest radio object, is 12,000 light-years away. Cygnus A is a pair of colliding galaxies 1000 million light-years from Earth!

The point is that the simplest radio telescopes extend the user's range of observation beyond the limits of our galaxy and close to the very edge of the known universe. The finest professional optical telescopes cannot boast of such a range.

A second extragalactic source is *Virgo A*. This elliptical galaxy generates 10^{31} kW of radio power from a distance of about 40 million light-years! The great distance makes Virgo A the least powerful of all the radio sources described in this chapter.

Optical Astronomy Principles and Projects

Chapter 2 deals with modern astronomy in very general and largely theoretical terms. This chapter brings the amateur astronomer into the picture. The primary objective here is to describe some of the know-how and techniques that are invaluable to optical astronomers. Amateur radio astronomers will find the discussion valuable to their own work, and there is always the hope that some enterprising investigator will find a way to tailor some of the conventional procedures of optical astronomy to fit into the realm of radio astronomy.

CASUAL AMATEUR ASTRONOMY

Casual amateur astronomers, or *starviewers*, are mainly interested in stepping into their backyards on a clear night and learning the positions of a few constellations or the names of some of the brighter stars. Their attachment to astronomy might be more romantic than scientific, and the ancient lore of the stars might be more important to them than knowing all about celestial mechanics or the origins of the stars.

Casual astronomers often take special delight in observing eclipses, meteor showers, and special planetary configurations. They might attempt to take some direct photos of celestial events of interest to them, but they seldom bother to gather any data that could be of use to more serious amateur or professional astronomers.

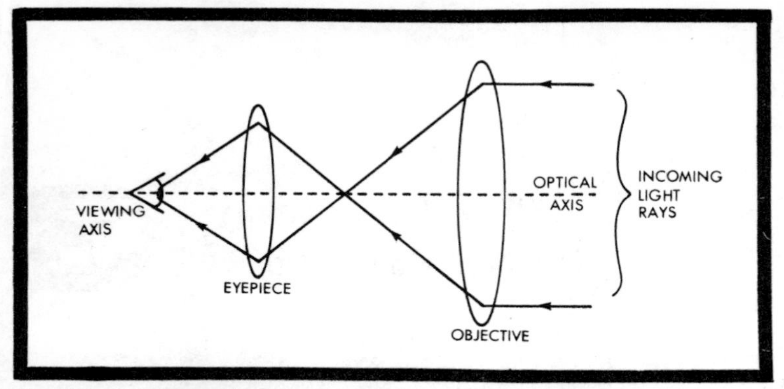

Fig. 3-1. Principle of a refracting telescope. Note that the viewing axis and optical axis are the same.

Backyard astronomers get most of the information from a wide variety of popular astronomy books written for young people and from amateur astronomy handbooks. Outdoor magazines sometimes carry announcements of special celestial events that can be appreciated without the aid of any special equipment, and dozens of major newspapers across the country carry regular columns geared to the interests and limited needs of starviewers.

OPTICAL TELESCOPES AND THEIR MOUNTINGS

Many backyard starviewers graduate to amateur astronomy when they acquire a top-of-the-line department-store telescope. Small telescopes of this variety can be purchased for prices ranging from about $15 to $100 or more, depending upon the size and quality of the lens system. The tripod mounting accounts for a good share of the cost of lower priced imported telescopes.

There are two basic types of optical telescopes: *refractors* and *reflectors*. As shown in Fig. 3-1, a refracting telescope uses a primary lens, called the *objective*, and a smaller lens system, called the *eyepiece*. The essential feature of a refracting telescope is that the viewing axis and the light-beam axis are one and the same.

As the name implies, a reflecting telescope (Fig. 3-2) uses a mirror and prism assembly to gather the light and reflect it out of the main housing to the observer's eyepiece. The mirror has a concave reflecting surface so that it can serve as the objective. Light entering the telescope as parallel rays is focused by the mirror onto the prism and eyepiece. The essential feaure of a reflecting telescope is that the viewing axis and light-beam axis are at a right angle to one another.

In principle at least, refracting telescopes might seem to be simpler than their reflecting counterparts. In practice, though, refracting telescopes are more expensive and tedious to build because of the high demands placed upon the quality of the lenses. Reflecting telescopes are relatively easy to build and they can work quite well at less-than-critical tolerances. For this reason, many serious amateur astronomers take special pride in building their own reflector telescopes—including the tedious procedure of grinding the mirror by hand.

Both refractor and reflector telescopes are available through department stores and science specialty houses such as Edmund Scientific Co. (See Figs. 3-3 and 3-4.)

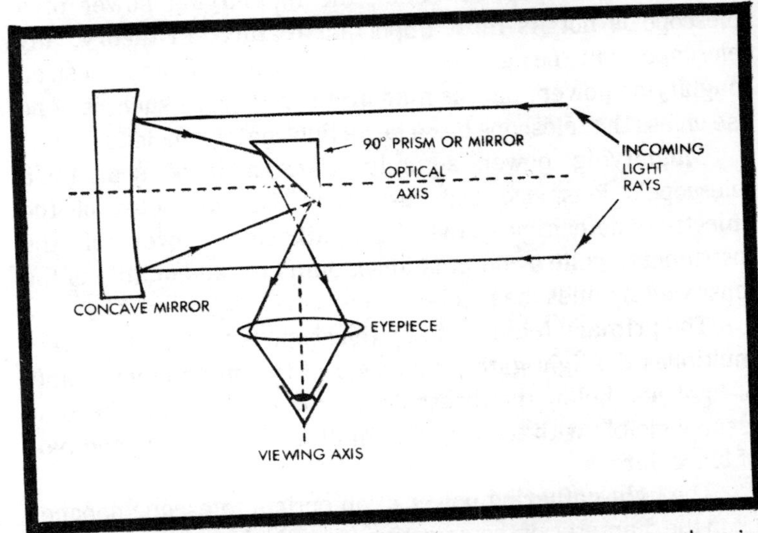

Fig. 3-2. Principle of a reflecting telescope. A concave mirror gathers incoming light rays and directs them to an eyepiece via a right-angle mirror or prism. The optical axis and viewing axis are at right angles to one another.

Fig. 3-3. A typical refracting telescope. The small telescope mounted on the main housing is a low-power refractor used for aligning the higher power optics. (Courtesy Edmund Scientific Co.)

Contrary to popular belief, the magnifying power of a telescope is not its most important feature. In theory, any telescope can be geared up to operate at any desired magnifying power; but the magnifying power, as such, is of no use unless the telescope has a large light-gathering lens.

Magnifying power actually takes a back seat to a telescope's lens size and resolution. The diameter of the objective determines the light-gathering power of the instrument, while a combination of diameter and quality of the lenses determines the resolving power.

The primary feature of any telescope is that it effectively multiplies the light-gathering power of the human eye. Points of light just below the threshold of human vision can become clearly visible with the aid of a small telescope or a good pair of binoculars.

The light-gathering power of an optical telescope depends upon the diameter of the objective lens: the larger the lens, the more stars the observer can see. The pupil of the human eye has an average diameter of about $^1/_3$ inch and can "resolve" about 500 stars on a reasonably clear night. A simple 2-inch telescope—one having an objective with a diameter of about 2 inches—increases the limits of optical sensitivity to a point where something on the order of 900,000 stars are observable. A 6-inch telescope increases the field to well over 5 million stars, and a good 8-inch model can bring into view more than 13 million stars.

One of the most fascinating views of the heavens I have ever seen was through a special homemade telescope having a magnifying power of 1—the lens had a diameter of about 12

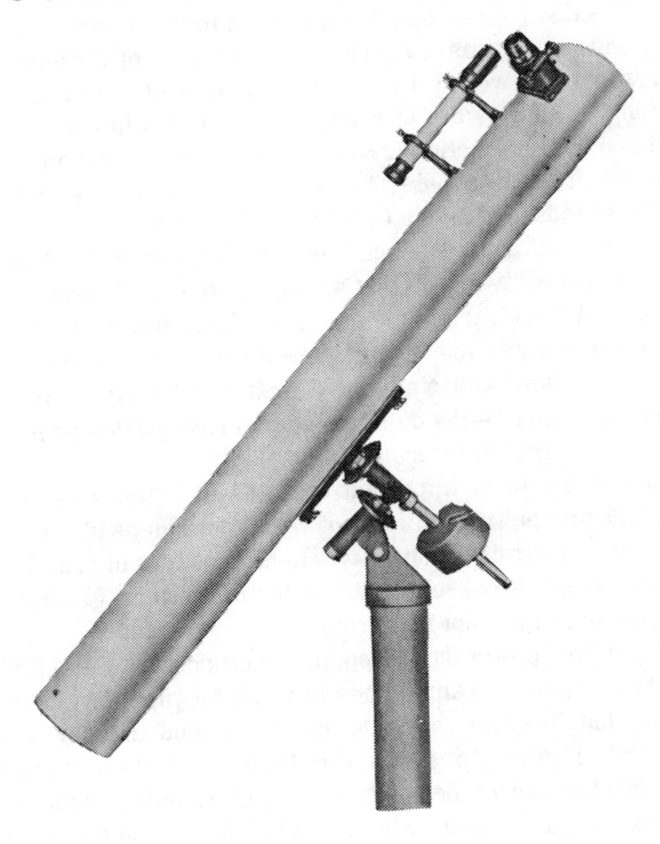

Fig. 3-4. A typical reflecting telescope. The eyepiece and a small spotting telescope are at the top of the main housing. The mounting for this telescope happens to be of the equatorial type. (Courtesy Edmund Scientific Co.)

inches, and the sight of millions of stars was nothing short of breathtaking! The point is that the light-gathering power of a telescope, determined by the diameter of the objective, is far more critical to satisfactory performance than simple magnifying power.

Besides increasing the light-gathering power of a telescope, the diameter of the objective determines the instrument's *resolution*. The unaided human eye can resolve two points of light that are no less than 180 seconds of arc apart. This distance is roughly equivalent to the diameter of a period on this page as it appears at a distance of three feet. The resolving power R (in seconds of arc) of a telescope, assuming the lenses are of reasonably good quality, may be calculated by the equation $R = 4.5/D$, where D is the diameter of the objective in inches. The number, 4.5, is the resolving ability (seconds of arc) of a 1-inch lens.

An observer using a 2-inch telescope, for example, should be able to resolve two stars that are no less than 2.25 seconds of arc apart. A 6-inch telescope has a resolving power of about 33.75 seconds, while the 200-inch telescope at Mt. Palomar in California's Cleveland National Forest can resolve objects 0.0225 second apart—*the diameter of a period on this page as viewed from a half-mile away*!

Two of the most important features of a telescope—its light-gathering power and limit of resolution—depend mainly upon the diameter of the objective. Lens quality is of practical importance too, because any discontinuities in the glass can produce image and color distortion.

Magnifying power isn't altogether unimportant, but a high magnifying power cannot possibly make up for lesser specifications as far as lens diameter and quality are concerned. Figure 3-5 shows the limits of light-gathering power as a function of the objective diameter, and the equation accompanying that figure can be used to determine the upper useful magnifying power as limited by the instrument's resolving power.

Most beginning amateur astronomers tend to believe the highest possible magnifying power is the best for all occasions. The preceding discussion should indicate that this is not the case at all. In fact, the "best" magnifying power is the lowest power that can do the job at hand. And in instances where the size and quality of the objective are inadequate, no amount of magnifying power can help.

As far as observing stars is concerned, the only reason for stepping up the telescope's magnifying power is to increase the apparent distance between them. Once the magnifying power exceeds 40 times the lens diameter, the images lose both their brightness and sharpness of focus. Increasing the magnifying power cannot produce the effect of bringing a star in closer, either. The stars are all so far away that even the 200-inch telescope shows them as points of light. There is no known technique for observing the disc of even the largest and nearest stars.

Higher magnifying powers are quite useful for making observations of objects in the solar system, nebula, and other

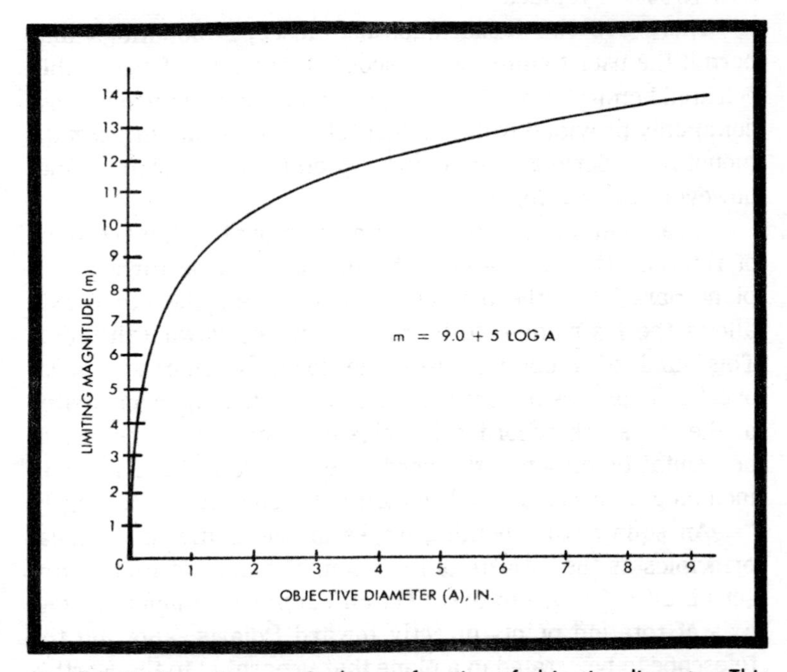

Fig. 3-5. Limiting star magnitude as a function of objective diameter. This curve shows that larger telescopes can resolve fainter stars than smaller telescopes can.

galaxies, however. As long as the user keeps the telescope within its limits of light sensitivity and resolution, he can magnify the images of such large and highly detailed objects as he chooses.

Even when used within the limits of the telescope, high magnifying powers can become a nuisance. At powers above 100× or so, the shimmering motion of the atmosphere tends to blur the images. And when using high-power lenses in a telescope mounted on inexpensive mountings, the slightest motion of the Earth and equipment can make the image flash completely out of the field of view.

Magnifying power, light-gathering power, and resolution all play important roles in the performance of an optical telescope. Novices tend to overplay the importance of magnifying power; and unfortunately, many are sold high-priced telescope kits merely because the set includes a useless 500× eyepiece.

There are two basic kinds of telescope mountings that permit the user to aim the telescope to any part of the visible celestial hemisphere. The simplest mounting and the one most commonly provided with low-cost telescopes is an *altiazimuth* mounting. Serious amateurs prefer the other type, however—an *equatorial* mounting.

An altiazimuth (altitude–azimuth) mounting has two axes of rotation. One axis allows the telescope to be turned in a plane parallel to the horizon (azimuth), and the other axis allows the instrument to be moved up and down (altitude). This kind of mounting uses the local horizon and local overhead point as a frame of reference. Since the coordinates of the stars are recorded in celestial coordinates instead of horizontal or altiazimuth coordinates, there is little point in including scales on a simple altiazimuth telescope mounting.

An equatorial mounting works on the same mechanical principles as the altiazimuth version. The axes of motion are set up differently, however. On an equatorial mounting, one axis of rotation points directly toward Polaris, allowing the telescope to be rotated in a plane that is parallel to the Earth's equator. The second axis of motion is perpendicular to the equatorial axis. The standard frame of reference for celestial coordinates is described in greater detail in the next chapter.

An equatorial mount always includes a set of scales that show the position of the telescope relative to the celestial sphere. One of the primary advantages of a scaled equatorial mount is that it lets the user pinpoint objects that are not visible with the naked eye. As long as the user knows the celestial coordinates of a stellar body, he can set the telescope to that part of the sky and find it in his field of view. It is possible to perform the same operation with an altiazimuth mounting, but the preliminary calculations can become rather cumbersome.

A special advantage of an equatorial mount is that it can track a celestial body across the sky by changing only one axis of motion. Any point on the celestial sphere moves across the sky in a path that is parallel to Earth's equator; and once the observer gets the object into the field of view, he can easily track it with slight motions on the equatorial axis. In the case of an altiazimuth mount, the observer has to change both the altitude and azimuth setting to keep the object in view.

It is possible—though impractical in most instances—to use an automatic *clock* drive with an altiazimuth mounting. Since an equatorial mount requires only one plane of motion to track an object across the sky, it requires only one drive motor to do the job automatically. A clock-driven altiazimuth mount requires a pair of drive motors synchronized in a complex fashion that, in effect, continuously compute the geometric difference between the horizon and equatorial frames of reference.

ASTROPHOTOGRAPHY

One of the most rewarding activities of amateur astronomy blends the skills of optical astronomy with those of photography. On the simplest level, it is possible to create some interesting star trails by simply pointing a camera to the sky and opening the lens for five minutes or so. On the most sophisticated level, attaching a camera to a telescope with a clock drive can increase the effective sensitivity of the instrument a hundredfold. Most professional optical astronomy is *astrophotography* today.

Photo emulsions have the ability to *integrate* incoming light rays. Whereas the human eye cannot both gather and store light energy, photo emulsions can. The longer a piece of film is exposed to the night sky, the more stars one sees on the film after it is processed. Even though a simple camera might not have an aperture much larger than the pupil of the eye, the film can integrate the light energy from very faint stars to the point where they become quite visible. The only limiting factor in the whole procedure is *fogging*—a gradual general exposure of the film caused by a "leaking in" of stray light.

Simple time-exposure photos of the sky produce what is commonly known as star-trail pictures. Star trails, as the Lick Observatory photo of Fig. 2-2 (preceding chapter) depicts, are arcs of light produced by the apparent circular motion of the celestial sphere. Polaris always appears as a single point of light on such a photo, while the other stars generate longer arcs. The farther a star stands from Polaris, by the way, the longer its trail will appear on the picture.

Project-minded amateurs sometimes produce their own star maps by taking a series of star-trail photos of the entire sky. By projecting the negatives onto a large screen, they can position the stars by making a dot at the beginning of each arc. A mosaic of such drawings can be assembled to produce a fine map of the heavens.

More serious astrophotographers attach a camera to the eyepiece of a telescope. A clock drive on an equatorial mount permits long time exposures of specific objects with little or no apparent motion appearing on the film.

DETERMINATION OF STAR BRIGHTNESS

Our own Sun is not especially bright as far as stars go; but to people living on Earth, it is the brightest object in the sky. Stars like Sirius might burn with optical intensities a million times greater than the Sun; but to people on Earth, the light from the mightiest stars seems puny by comparison.

The discrepancy between the actual brightness of a star and its apparent brightness as observed from Earth is largely due to the tremendous distances that separate the stars from the observer. Unless an astronomer happens to be doing some work with stellar physics, he has little interest in the true brightness of stars. For the most part, amateur astronomers are more concerned with the stars' brightness as observed from Earth.

Astronomers have devised a formal scale for rating the apparent brightness of stars. This magnitude scale assigns numbers to the stars in a fashion that shows larger numbers for fainter stars. Fainter stars, in other words, have larger magnitude ratings than brighter ones do.

To get a feeling for how the magnitude scale works, consider the fact that most of the stars in the Big Dipper portion of *Ursa Major* have a magnitude rating of 2. The faintest star an observer can see on a very dark and clear night with the naked eye is on the order of magnitude 6.

One particular star, some of the planets, and the Sun and Moon are so much brighter than the rest of the stars that the magnitude scale has to be extended through zero and into negative numbers. The Sun, for example, has a magnitude rating of −26.7, while Sirius, the brightest star ever seen from anywhere on the Earth, is of magnitude −1.6.

Amateur astronomers estimate the magnitude of a star by comparing its brightness with several others of known magnitude. Experienced observers can estimate stellar magnitudes within 10% of the figures found by means of elaborate optical measurements.

Appendix A shows the magnitude ratings of a few of the brighter stars visible from the northern hemisphere.

VARIABLE STAR OBSERVATIONS

Amateur astronomers wanting to develop their hobby to a point where they can make significant contributions to professional astronomy can engage in projects such as surveying *variable stars*. Variable stars are those which for one reason or another change brightness on a more or less regular schedule. The cycle can be as short as every three or four days for some stars and as long as thirty or forty years for others.

The reasons for the changes in the brightness of variable stars are not known in all instances; but it is known that many variable stars are *eclipsing binaries*—pairs of closely spaced stars that orbit around one another. When the two stars are separated from one another as viewed from Earth, they have the appearance of a single star of increased brightness. As one of the stars moves in front of its companion, the "star" seems to dim.

Optical astronomers have been able to distinguish one star from the other in a number of different eclipsing binary systems. In other cases, however, the stars are so far away or so close together that it is impossible to see them as distinct entities.

Amateur astronomers interested in observing the changes in intensity of variable stars can join the *American Association of Variable Star Observers* (AAVSO). This old and well established organization supplies qualified amateurs with highly detailed star maps and computerized prediction tables. The observer should estimate the brightness of the stars assigned to him over a certain period of time, and report his results to the AAVSO. The AAVSO uses the information to update its precision charts. The idea is to continually and progressively add to the computer's memory in the hope of uncovering "new" facts about the stars.

Beginners wanting to try their hand at observing some variable stars can find an abbreviated prediction chart in every issue of *Sky and Telescope*.

LUNAR OCCULTATION OBSERVATIONS

The Moon moves across the sky a bit more slowly than the stars do. As a result, stars frequently seem to disappear behind the eastern edge of the Moon and reappear on the western edge at some later time. Whenever the Moon hides a star in this fashion, it is said to *occult* that star. The study of lunar occultations over the years has led to a better understanding of the positions of the stars and the orbits of the Earth and Moon.

One of the most dramatic and scientifically useful kinds of lunar occultations occurs when a star barely skims across the lunar disc. As the star moves between high mountain peaks on the Moon, its light seems to flash off and on. And by making a number of simultaneous observations over a large geographical area, it is possible to draw a profile of the lunar landscape in rather striking detal.

These *grazing* occultations take place about twice a year in any given area of the Earth, and *Sky and Telescope* usually carries the appropriate prediction charts for interested amateurs.

The procedures for timing lunar occultations are rather simple, but the precision can be on the order of 0.1 second of arc or better. Amateur radio astronomers should study the occultation timing procedures with special interest because the techniques are readily adaptable to similar kinds of studies in radio astronomy.

One technique for timing occultations uses a portable tape recorder and a shortwave radio tuned to WWV (National Bureau of Standards official time station) or the Canadian time station, CHU. Just before the occultation is scheduled to begin, the observer sights the star in his telescope and starts the tape recorder. The instant the star disappears behind the lunar disc, he says something like, "Mark—disappear." And when the star reappears, he says, "Mark—reappear." This information, along with the radio time signals, are recorded on the tape; and by playing the tape at a slower speed at some later time, the observer can use an ordinary stopwatch to determine the elapsed time between a radio time "tick" and his "mark" signal that follows it. Adding this elasped time to the radio time provide occultation time information that is accurate within the observer's reaction time—usually about 100 milliseconds (0.1 second).

OBSERVATIONS OF METEORS

Meteors or *falling stars* are sand-like bits of space debris that become trapped in the Earth's gravitational field. As they enter the upper atmosphere at tremendous speeds, the resulting friction of stationary air particles on the plummeting debris generates enough heat to vaporize the material. Most meteors vaporize completely before they ever have a chance to reach the Earth's surface. They are nevertheless fascinating objects because of the brief, blue-white trails of molten rock they leave behind.

The space debris—the stuff of meteors—rests in about six large pools in the Earth's orbit. Meteor showers thus tend to occur at about the same time each year, and they take on the names of the constellations that appear to mark the origin of the trails. The constellation of Perseus, for example, appears to be the point of origin for meteor trails belonging to the annual Perseid shower (August 10–14).

Meteor trails appear in the sky throughout every night during a scheduled shower. The show is always most intense, however, during the predawn hours when the observer's point on Earth is plunging head-on into the pool of space debris.

Many amateur astronomers like to count meteor trails during a scheduled shower. A more interesting and demanding project calls for plotting the length, direction, and time of occurrence of each trail on a star map. A pair of observers located a mile or so apart can compare their results and use the process of triangulation to determine the meteors' altitudes and velocities through the atmosphere.

Meteors have been known to produce some strange whistling sounds in amateur radio equipment. Radar studies of meteors have proved to be quite fruitful, but passive radio studies are much less reliable.

UNIVERSAL AND LOCAL TIME

Just about every kind of observation in astronomy requires a knowledge of time. The Sun, Moon, planets, and stars all rise and set at certain times. And since these objects move continuously across the sky, the observer has to apply time corrections and equations to determine their positions.

Much of the routine work in astronomy is concerned with compiling and interpreting prediction tables; and even a cursory examination of astronomical tables in popular almanacs demonstrates the importance of timekeeping. Everything in the universe is in motion with respect to the Earth, and knowing when and where an event is to take place is pretty much a matter of knowing the times involved.

Radio amateurs should be quite familiar with the terrestrial time standard known as *Greenwich Mean Time* (GMT). GMT, normally expressed in 24-hour time, is the local time in the London borough of Greenwich—a small city that lies directly on the Earth's *prime meridian* (0° longitude).

Astronomers throughout the world have adopted a different name for GMT. The astronomical expression is *Universal Time* (UT); but for all practical purposes, GMT and UT are one and the same. When it is 11:30 p.m. in Greenwich, for instance, GMT and UT are both expressed as 2330 hours or 23^h30^m (23 hours, 30 minutes from midnight last, or time zero).

It would be quite convenient for astronomers if everyone used the UT standard. There is no real reason why all clocks in the world cannot be set to a single time standard. The only trouble is that people like to have midnight fall at nighttime and noon occur in the middle of the day. This arbitrary notion imposes the concept of *Local Time*(LT).

The world is divided into 25 more or less equal time zones. The lines separating these LT zones tend to follow geographical longitude lines at 15° intervals. The LT zone lines are bent so that they never pass through highly populated regions of the world. If this were not the case, one side of a town could be running on a schedule that is an hour away from the other side.

The LT in any zone is one hour different from the LT in the zones on either side of it. The time change is such that the time zone to the west is one hour behind and the zone to the east is one hour ahead. Table 3-1 shows the LT zones for the continental U.S. The table shows that an observer working in the CST zone, for example, will read 3:00 p.m. on his clock when it is 2:00 p.m. MST and 4:00 p.m. EST.

Table 3-1 also shows the corrections required for converting UT to LT. To find EST from UT, for instance, it is necessary to subtract 5 hours from UT. And if an eclipse is scheduled to begin at 06^h20^m UT, an observer in New York City knows he can see the eclipse begin at $06^h20^m - 5^h = 01^h20^m$ or 1:20 a.m. EST.

Daylight Saving Time (DST) complicates the conversion procedure only a little bit. When DST is in effect, it is

Table 3-1. Time Correction Factors for U.S. Time Zones

Time Zone	Abbv.	Central Meridian	Time Correction
Atlantic Standard	AST	60°W	4^h
Eastern Standard	EST	75°W	5^h
Central Standard	CST	90°W	6^h
Mountain Standard	MST	105°W	7^h
Pacific Standard	PST	120°W	8^h

necessary to add one more hour to the conversion result. If the observer in New York is planning to observe the same eclipse in Table 3-1, and it is Daylight Saving Time in his area, he should do the following:

$$
\begin{array}{rl}
06^h20^m & \text{UT} \\
-\underline{05^h00^m} & \text{TC} \\
01^h20^m & \text{EST} \\
+\underline{01^h00^m} & \text{DST correction} \\
02^h20^m & \text{EDST}
\end{array}
$$

The procedures for converting LT to UT are just the reverse of the UT-to-LT procedure. If an observer notes an event taking place at 10 p.m. EST on April 3, he can convert to UT by adding 5 hours:

$$
\begin{array}{rl}
22^h00^m & \text{EST April 3} \\
+\underline{05^h00^m} & \text{TC} \\
27^h00^m & \text{UT}
\end{array}
$$

Since there are only 24 hours in a day, the result of 27^h must be interpreted as meaning the event will take place 3 hours into the next day. Continuing the calculation:

$$
\begin{array}{rl}
27^h00^m & \\
-\underline{24^h00^m} & \\
03^h00^m & \text{UT April 4}
\end{array}
$$

The UT/LT corrections described thus far are wholly adequate for determining the time of occurrence of events taking place simultaneously everywhere on Earth. Such events include equinoxes, solar flares, eclipses of Jupiter's moons, and any other event that is not directly associated with the Earth's rotation.

The time of occurrence of events that are directly associated with the Earth's rotation—the rising and setting times of celestial bodies, for instance—can only be partly determined by the application of the usual UT/LT corrections.

Each time zone represents about 15° of longitude or about 1 hour of rotation time for the Earth. If the Sun is scheduled to rise at a certain time in a given time zone, it actually rises 1 hour sooner at the eastern edge of the zone than it does on the western edge. Civil standard times are set to the middle of each time zone, so a simple UT/LT conversion could be off as much as 30 minutes in either direction when it comes to rotation-dependent events. The error can be even more significant in instances where the time zone is more than 15° across.

The precision and accuracy of amateur radio astronomy equipment is rather sloppy compared to the optical counterparts. Even so, an error of 30 minutes or more is quite significant; and knowing how to make exact time corrections for rotation-dependent events is important to nearly all radio astronomy work.

The Earth rotates approximately 360° for every 24 hours that pass. This means the Earth rotates $360°/24^h$ or 15° per hour—a fact that explains why civil time zones are about 15° wide. Taking this line of reasoning a bit further, it turns out that the Earth rotates 0.25°/min; or to state it another way, the Earth requires 4 minutes to rotate 1°.

If the Sun is thus scheduled to rise at 6:00 a.m. at one point on the Earth, it will actually rise at 6:04 a.m. at another point 1° to the west. Conversely, the sun will rise 4 minutes sooner at a point 1° to the east. To cite a more specific example, suppose an observer is located at 84°W longitude. He is, in other words, 84° west of Greenwich. Any rotation-dependent event that is scheduled to take place at a certain time in Greenwich will take place at 84°W longitude $4 \times 84 = 336^m$, or 5^h36^m later. If a certain star is supposed to rise at 0500 UT in Greenwich, it will rise 5^h36^m later at 84°W longitude.

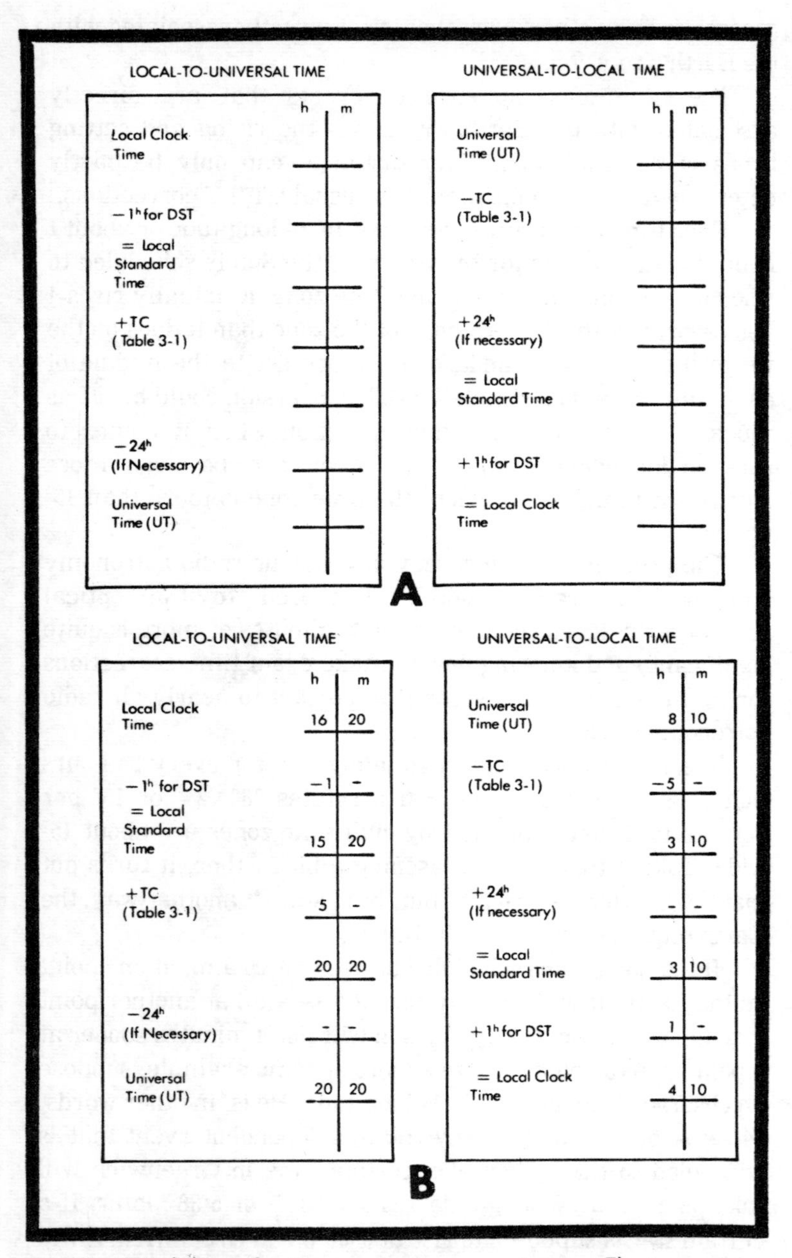

Fig. 3-6. Worksheets for time-conversion computations. The two entries at the bottom show worked-out examples.

The product of 4 min/degree and the local longitude is called the *Longitude Time Correction* (LTC) throughout this book. For points west of Greenwich, the LTC is added to UT; and for points to the east of Greenwich, the LTC has to be subtracted.

It is most important to note that applying an LTC to UT provides another UT, and not a Local Time. The star that rises at 05^h00^m UT in Greenwich rises 5^h36^m later at 84°W longitude. The LTC in this example is 5^h36^m, and adding that figure to the given UT results in 10^h36^m UT. This means the star rises at 84°W longitude when the clocks in Greenwich read 10:36 a.m. The 10^h36^m figure *does not* represent the local time of occurrence!

To determine the local time at which the star rises, the observer has to perform the UT/LT correction described earlier in this section. If the star is to rise at 10^h36^m UT in his area, the LT of the event is:

$$
\begin{array}{rl}
10^h36^m & \text{UT} \\
-05^h00^m & \text{TC} \\
\hline
05^h36^m & \text{LT or 5:36 a.m. EST.}
\end{array}
$$

Figure 3-6 shows some sample computation sheets that can simplify the conversion process.

Some Useful Coordinate Systems

Astronomers use several different sets of coordinate systems to specify the positions of optical and radio sources in the sky. Any amateur radio astronomer who intends to carry out meaningful projects must become familiar with the *coordinate systems* and learn to apply them under the appropriate circumstances.

The simple star maps shown in Chapter 2 can indeed give the user some idea of the positions of stars on a given set of dates and times, but such maps lack the precision and accuracy that is inherent in the more formal coordinate-system charts described in this chapter. Even in instances where the precision of an amateur's equipment far exceeds that of the formal charts, it is important to understand and use the language of the coordinate systems when communicating ideas to other astronomers.

This chapter deals with three coordinate systems. Two of them are *geocentric*—they use Earth as a frame of reference. The third system uses the celestial sphere as a frame of reference.

GEOGRAPHICAL COORDINATES

Most would-be amateur radio astronomers are quite aware of the fact that it is possible to specify the position of any point on the surface of the Earth in terms of its latitude and longitude. Reviewing the principles of geographical coordinates might thus seem to be a waste of time. A brief review of latitude and longitude is in order, however, because the correlation between this familiar system of coordinates and the lesser-known horizon and celestial coordinate systems is a very close one in many respects. A brief review of the familiar system can lead to an easier understanding of the others.

Geographical Longitude

Geographical longitude is measured with reference to great circles that run in a north–south direction and intersect at the Earth's poles. As shown in Fig. 4-1, the longitude circle that passes through Greenwich is always taken as the origin for longitude measurements. This particular longitude *great circle* is known as the *prime meridian.*

Measurements in a westerly direction from the prime meridian (0° longitude) run through a full 180° to a point halfway around the Earth to the mid-Pacific. Longitude 75°W, for example, passes between New York City and Washington, D.C., while 120°W longitude passes just west of Santa Barbara, California.

Longitude measurements in an easterly direction run from 0° at Greenwich to the same 180° line in the Pacific. Longitude

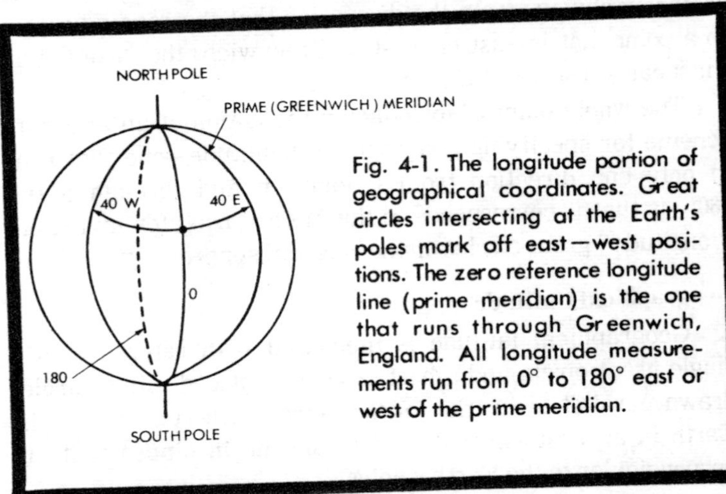

Fig. 4-1. The longitude portion of geographical coordinates. Great circles intersecting at the Earth's poles mark off east—west positions. The zero reference longitude line (prime meridian) is the one that runs through Greenwich, England. All longitude measurements run from 0° to 180° east or west of the prime meridian.

75°E passes between Delhi and Bombay, India, while longitude 120°E runs very close to Manila in the Philippines.

The longitude great circles indicate the east–west angular distance between Greenwich and any other point on Earth. The precision of a geographical map depends upon the number of longitude great circles inscribed on it. If a map is divided by 36 equally spaced longitude lines, the precision is no better than 10°. If, on the other hand, the map is divided into 360 longitude lines, the map's error tolerance is on the order of 1° of arc. Each degree can be divided into 60 minutes of arc, and each minute can, in turn, be divided into 60 seconds of arc. County surveyor maps show a limited geographical area subdivided into degrees, minutes, seconds, and even decimal parts of seconds.

In principle at least, there is no reason for measuring longitude from 0° to 180° in both the east and west directions. The whole system could be revised to run from 0° through a full 360° in one direction—say, the westerly direction. New York City and Santa Barbara would still be at longitude 75° and 120°, respectively; but the longitudes of Delhi and Manila would be specified at 285° and 240°, rather than 70°E and 120°E.

As the geographical longitude system stands now, however, no reading can be greater than 180°; and unless it is somehow obvious from the discussion that the reading refers to a point that is east or west of Greenwich, the designation must carry a suffix of E or W.

The whole point of introducing the notion of an alternate scheme for specifying geographical longitude—one that runs in only one direction from Greenwich and through a full 360°—is that it corresponds to the method used for laying out "longitude" great circles on the celestial sphere.

Geographical Latitude

Geographical latitude is measured with reference to a single great circle and a host of evenly spaced minor circles drawn parallel to it. The *latitude* circles all run around the Earth in an east–west direction, and lie in planes that are perpendicular to the Earth's poles.

The single-latitude great circle is, of course, the Earth's equator. As illustrated in Fig. 4-2, the equator is always taken as the zero reference circle for measuring latitude degrees. Latitude measurements begin with the equator at 0° and run through 90° of arc in the north and south directions. The north pole, for example, is at 90°N latitude, while the south pole is at 90°S latitude. Columbus, Ohio, is very close to 40°N latitude and Wellington, New Zealand is at about 40°S latitude.

The latitude great circles indicate the north–south angular distance between the equator and any other point on Earth. As in the case of longitude markings, the precision of a geographical map depends upon the "fineness" of the latitude markings. Latitude, like longitude, can be broken down into minutes, seconds, and decimal parts of seconds of arc.

The idea of setting the equator as a zero reference circle for latitude and measuring angular distances northward and southward 90° works out quite well. And as long as the latitude designation for a point on Earth is followed by an N or S, there can be no confusion about whether the point is in the northern or southern hemisphere.

It could be just as useful, however, to prefix all northern latitude readings with a plus sign (+) and all southern readings with a minus sign (–). Instead of designating the latitude of Columbus, Ohio as 40°N, it could well be written as

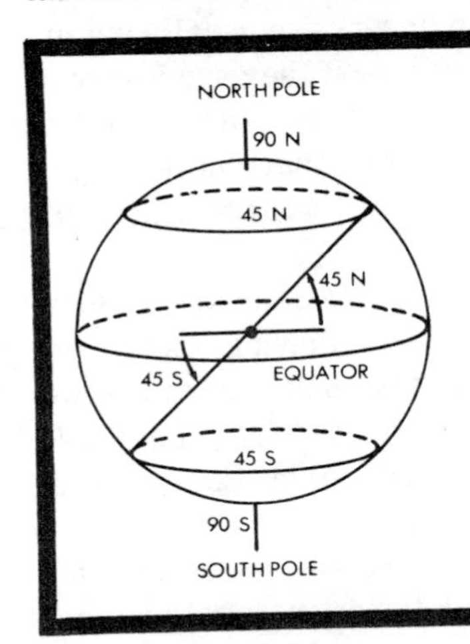

Fig. 4-2. The latitude portion of geographical coordinates. The Earth's equator is taken as the zero-reference circle for marking off geographical latitude. Circles parallel to the equator indicate latitude from 0° to 90° north and south.

+40°. The latitude of Wellington, N.Z., would thus be written as −40° rather than 40°S.

The conventional N and S method of designating geographical latitude is no less meaningful than a (+) and (−) counterpart. The point of suggesting the (+) and (−) scheme for latitude is that it corresponds directly to the method used for designating latitude on the celestial sphere.

Determining Latitude and Longitude

The procedures for accurately determining the latitude and longitude of a point on the Earth can be a rather complex one that is better handled by skilled surveyors and navigators. An amateur radio astronomer must have a fairly accurate notion of the latitude and longitude of his observatory, but there is no need to determine it for himself from scratch.

Standard world atlases and globes show the geographical coordinates of points within 10° or so. In many instances, the maps include an interpolation scale that lets the user figure the coordinates down to about 1°.

The popular newsstand almanacs give the latitude and longitude of major cities to the nearest second of arc. Experimenters who do not live in one of the designated cities have to resort to another means of getting the information.

The simplest way to obtain the most accurate figures for latitude and longitude is through the services of the local county engineer's office. County engineers are usually willing to provide the information free of charge if they are aware of why the inquiry is coming to them in the first place. A letter, telephone call, or personal visit is about all that's necessary to determine the position of a radio observatory to the nearest 0.1 second of arc.

To illustrate the various procedures for determining local longitude and latitude, consider the following examples: According to a standard geography textbook, my own radio antennas in Columbus, Ohio, are at 83°W 40°N. The *World Almanac* lists the latitude and longitude of Columbus at 83°00′17″W and 39°57′47″N.

The information from the *World Almanac* is actually more accurate than any kind of calculations an amateur may require. I've rounded off the figures to the nearest degree and found them to be satisfactory for all routine work. In the example above, then, the best all-around figures for my own latitude and longitude are 83°W, 40°N.

CELESTIAL COORDINATES

At first glance the coordinates on a celestial sphere seem identical to the latitude and longitude markings on a globe of the Earth. The markings are in fact the same—only the nomenclature and the methods of measuring the angles are different. (See Fig. 4-3.)

Before describing celestial latitude and longitude in detail, it is important to define a few terms. The opening paragraphs of Chapter 2 describe the general appearance and purpose of a celestial globe. What is important at this point is the fact that the celestial sphere has poles and an equator that correspond to Earth's poles and equator. The poles and equator on the celestial sphere have the same names as the corresponding elements on the Earth. Terms such as *celestial north pole*, *celestial south pole*, and *celestial equator* should not cause any particular confusion.

Differences between the Earth's coordinate system and that of the celestial sphere enter the scene when considering geographical and celestial latitude and longitude. Rather than

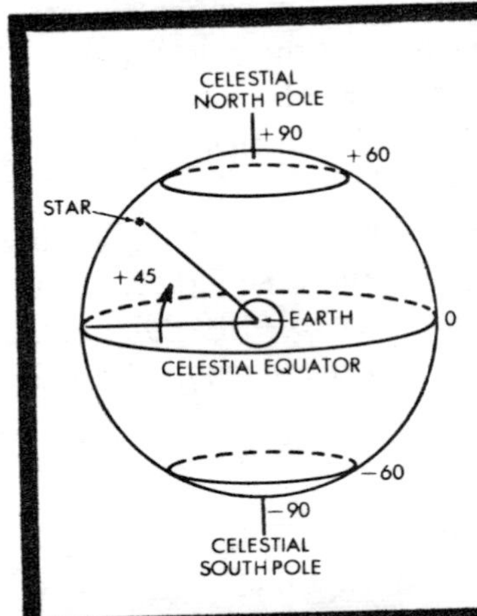

Fig. 4-3. The declination portion of geographical coordinates. The imaginary celestial sphere uses declination measurements to mark off angular distances between the celestial equator and the celestial north and south poles. Being similar to geographical latitude, declination is measured from 0° at the celestial equator to 90° plus or minus at the north and south celestial poles, respectively.

specifying the position of a point on the celestial sphere in terms of N and S latitude and E and W longitude, the corresponding terms in celestial coordinates are + and − *declination* and *right ascension.* Declination is measured in degrees from the equator just as latitude is on the Earth. Right ascension, however, is measured in terms of hours and minutes of time from 0^h to 24^h, rather than E and W longitude.

Declination

Declination is to celestial coordinates as latitude is to geographical coordinates. Declination, in other words, is measured with reference to a single great circle and a set of minor circles parallel to it. The declination circles all lie in planes that are perpendicular to the axis of the celestial sphere.

As shown in Fig. 4-3, the single great circle of declination is the celestial equator; and this circle is always taken as the zero point for declination measurements. Declination measurements run from 0° at the celestial equator through 90° to the north celestial pole and 90° south to the south celestial pole. Declination readings in the northern celestial hemisphere are prefixed with a plus sign (+), while readings of declination in the southern celestial hemisphere are prefixed with a minus sign (−).

A star having a declination of +45°, for example, rests halfway between the celestial equator and the celestial north pole. A star having a declination of −45° also rests halfway between the celestial equator and one of the poles; but in this instance the pole is the southern one.

Any point on the celestial sphere having a declination of +90° rests on the celestial north pole. And it so happens that Polaris has a declination that is very close to +90°. Polaris, in other words, marks the celestial north pole for most practical purposes.

Right Ascension

The right ascension of a point on the celestial sphere corresponds to the longitude angle of a point on the surface of Earth. Right ascension is measured with reference to a series of great circles, properly called *hour circles*, that intersect at the celestial poles. (See Fig. 4-4.) One of the hour circles must be chosen as the zero reference circle and serve the same general function as the prime meridian on the geographical coordinate system.

The zero-reference hour circle, or celestial prime meridian, is the hour-circle position of the Sun at the moment of the March equinox. Right ascension (RA) is measured off from this 0° celestial meridian eastward through a full 360°. An object on an hour circle that is 90° east of the celestial prime meridian, for example, would stand one-fourth of the way around the celestial sphere from the celestial prime meridian. An object on the 355° hour circle east of the celestial prime meridian would be just 5° west of that zero-reference hour circle.

Right ascension expresses the angular distance between the zero-reference hour circle and the hour circle of a star. Just as longitude degrees can be used to express the "vertical" great-circle position of a point on the Earth, right ascension expresses the vertical great-circle position of a star—and all stars are assigned right-ascension coordinates.

The whole scheme of right-ascension coordinates is not based upon measures of degrees, however. Instead, right ascension is expressed in hours, minutes, and seconds of time.

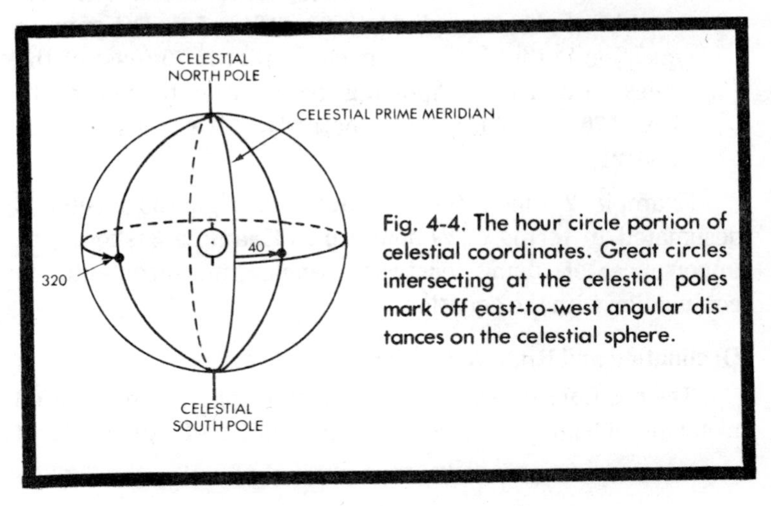

Fig. 4-4. The hour circle portion of celestial coordinates. Great circles intersecting at the celestial poles mark off east-to-west angular distances on the celestial sphere.

The reasoning behind the hour designation of right ascension is the fact that the celestial sphere rotates through a full 360° in one *sidereal day.** Since there are 24 sidereal hours in one sidereal day, every 15° change in the position of the celestial sphere represents the passage of one hour of sidereal time. A point resting on an hour circle that is 30° east of the celestial meridian has a right ascension of $30°/15 = 2^h$.

The following equation is used for translating angular celestial distances into hours of right ascension:

$$RA = \frac{\text{eastward angle from celestial meridian}}{15} \qquad (4\text{-}1)$$

Carefully studying the following examples can lead to a better understanding of right ascension and the application of the equation.

Example 1: What is the right ascension of a point having an hour-circle position of 45° east of the celestial meridian? According to the equation, $RA = 45/15 = 3^h$.

Example 2: What is the right ascension of a point on an hour circle that is 335° east of the celestial meridian? According to the equation, $RA = 225/15 = 15^h$.

Example 3: What is the right ascension of a point on an hour circle that is 290° *west* of the celestial meridian? Since right ascension is measured eastward, the 290° west has to be converted to $360° - 290° = 70°$. The point in question is thus on an hour circle that is 70° east of the celestial meridian. Applying the equation to this result, $RA = 70/15 = 4.67^h$. Alternatively, $RA = 4^h40.2^m$ or $4^h40^m21^s$.

Example 2 illustrates the fact that right ascension is designated in terms of 24-hour time. Example 3 shows the importance of using eastward angle measurements in conjunction with the equation.

Declination and Right Ascension

The positions of points on the celestial sphere are specified in terms of right ascension and declination. In principle at

* A sidereal day is equal to $23^h56^m03^s$ of ordinary clock time.

least, right ascension corresponds to geographical longitude and declination corresponds to latitude. (See Fig. 4-5.)

Points lying directly on the celestial equator have a declination of 0°. The right ascension of these points, however, depends upon their angular distance measured eastward from the 0° hour circle. Recall that the Sun crosses the 0° hour circle at the time of the March equinox. At the instant of the March equinox, the Sun is also crossing the celestial equator. The special point in the sky that marks the position of the Sun during the March equinox thus has celestial coordinates of 0° declination and 00^h RA. The particular point in the sky serves as the zero reference point for the entire celestial coordinate system; and because of its special significance, it takes the special name, *First Point of Aries*.

Considering a few more examples of celestial coordinates might be instructive at this point. The coordinates of Polaris are +89°08′ and $2^h05^m25^s$. Note that the declination of Polaris is rather close to +90°, placing it almost directly on the celestial north pole. Sirius, the brightest star in the sky, has a declination of −16°41′ and a right ascension of 6^h44^m. Sirius is

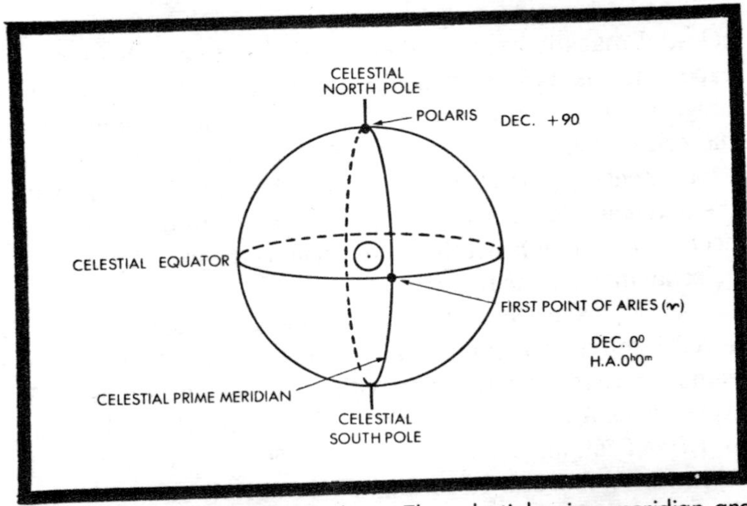

Fig. 4-5. The basic celestial sphere. The celestial prime meridian and equator intersect at a point in the sky that is known as the **First Point of Aries.** The First Point of Aries is taken as the zero-reference point for the entire system of celestial coordinates.

thus located about 17° below the celestial equator and about 100° east of the 0° hour circle.

The celestial maps in Figs. 4-6 through 4-11 show a strip of the celestial sphere that runs through a full 24 hours of right ascension and between −70° and +90° of declination. These celestial maps are quite different from the simple star maps shown in Chapter 2. Whereas the simple star maps show the night sky as it actually appears on certain hours of certain nights, the celestial maps in this chapter are much more universal in their application. The celestial maps are more accurate and precise, too. Distortions inevitably introduced into two-dimensional representations of a spherical surface are not really relevant as far as the celestial maps are concerned. One of the primary applications of a celestial map is to help an astronomer determine the celestial coordinates of any point on the celestial sphere—it is up to him to locate that particular point in the sky as he sees it.

In the celestial maps, constellations are abbreviated with a three-letter identifier; Appendix E lists the constellations and their abbreviations.

The best way to gain an appreciation of the usefulness of celestial maps is by using them. The time to get this kind of experience is before launching a formal radio astronomy program, however; and a few moments spent learning to use the celestial maps will pay off when the results of the first experiments begin to take shape. For a beginner, working with the antenna and receiver and gathering useful data can be a hectic process—it is not the time to double the work by facing a celestial map for the first time, too.

As an exercise, estimate the celestial coordinates of several major optical or radio sources on the maps and compare your own findings with the coordinates listed in Appendixes A and B. The map in Fig. 4-6, for example, show α Eridani at about 1^h36^m RA and −50° dec. According to Appendix A, that star's coordinates are 1^h37^m RA and −57° dec. The estimate from the map compares favorably with the star's actual coordinates. The slight error is well within the tolerance of most amateur radio astronomy equipment.

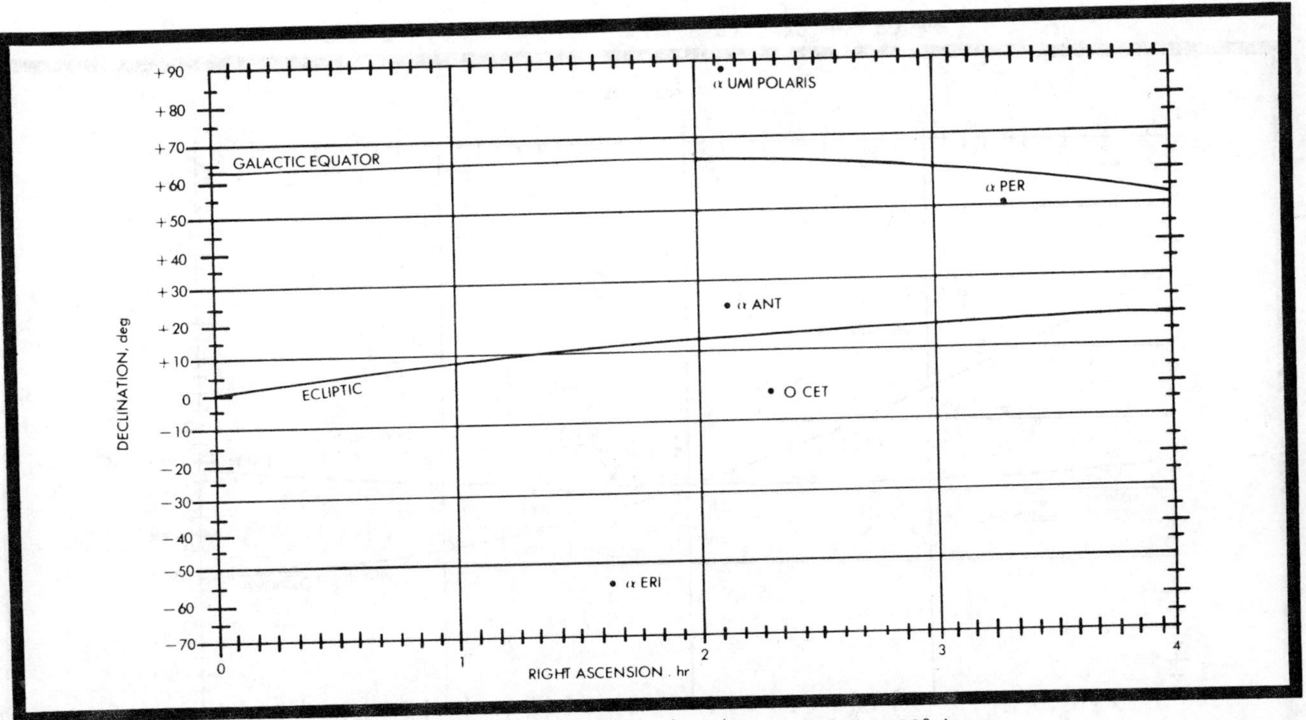

Fig. 4-6. A basic celestial map—0^h to 4^h RA, $-70°$ to $+90°$ dec.

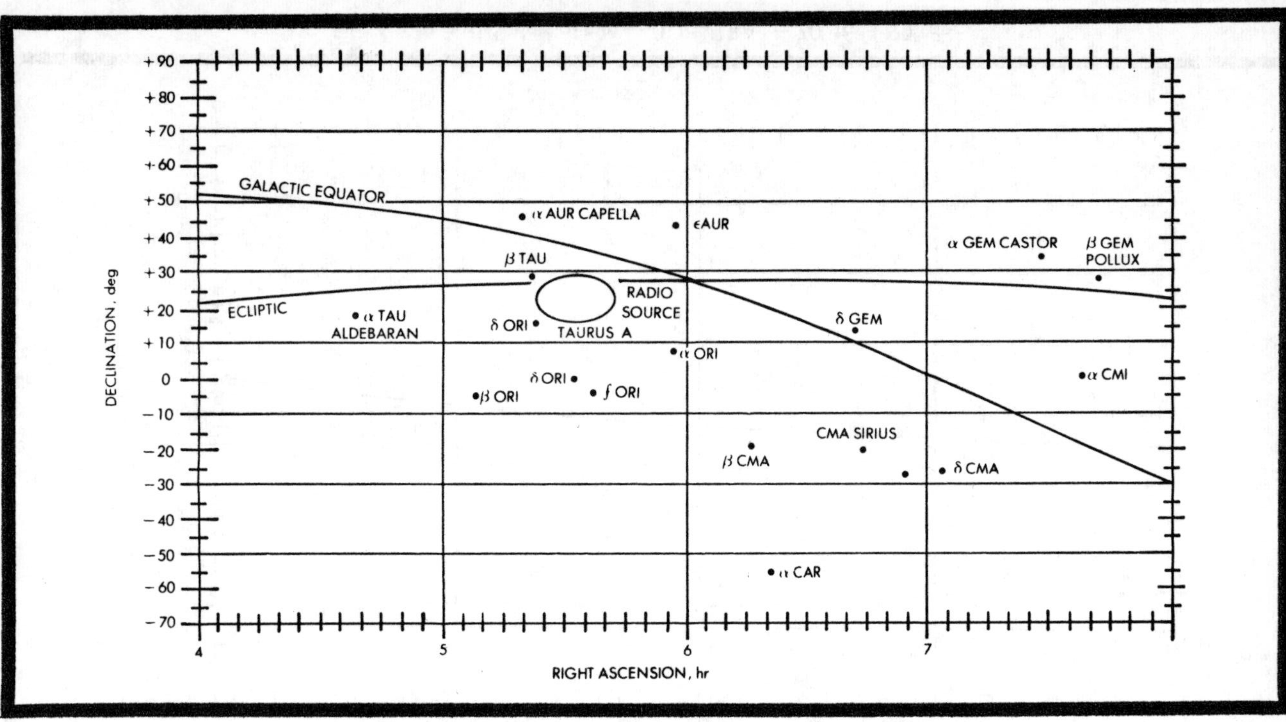

Fig. 4-7. A basic celestial map—4^h to 8^h RA, −70° to +90° dec.

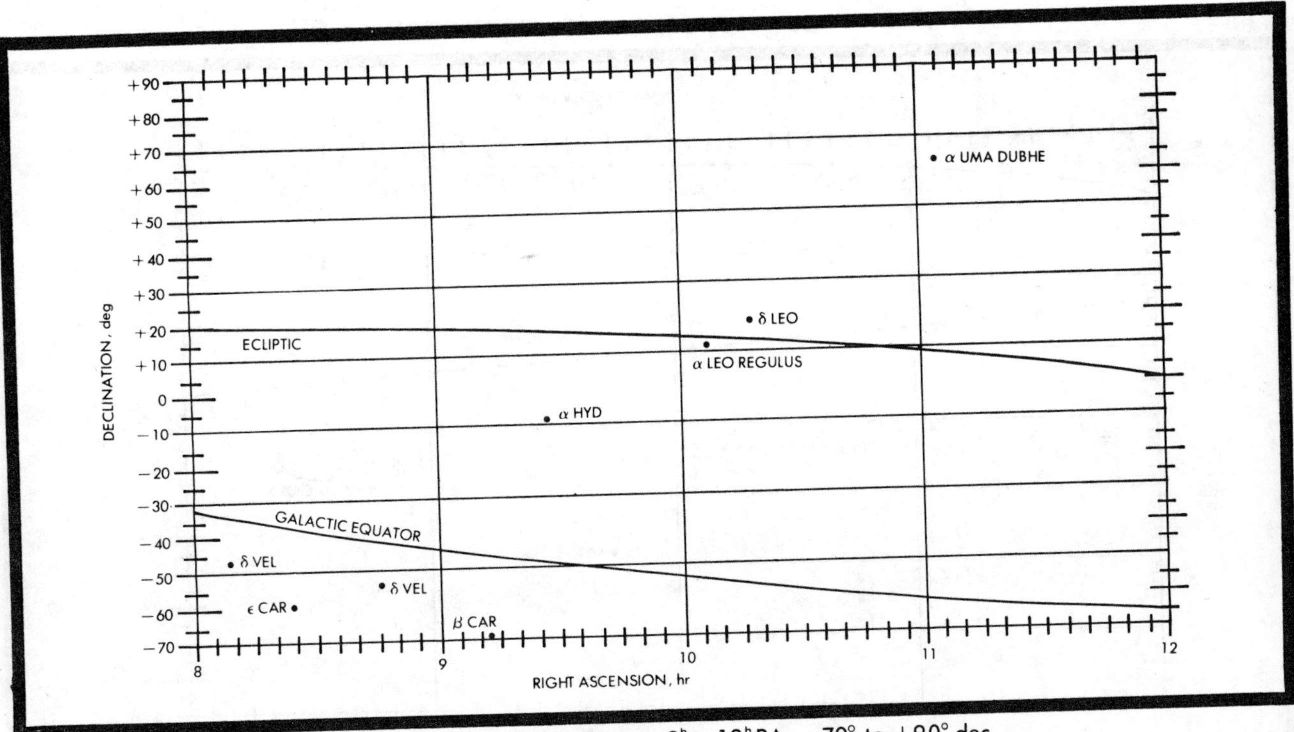

Fig. 4-8. A basic celestial map—8^h to 12^h RA, −70° to +90° dec.

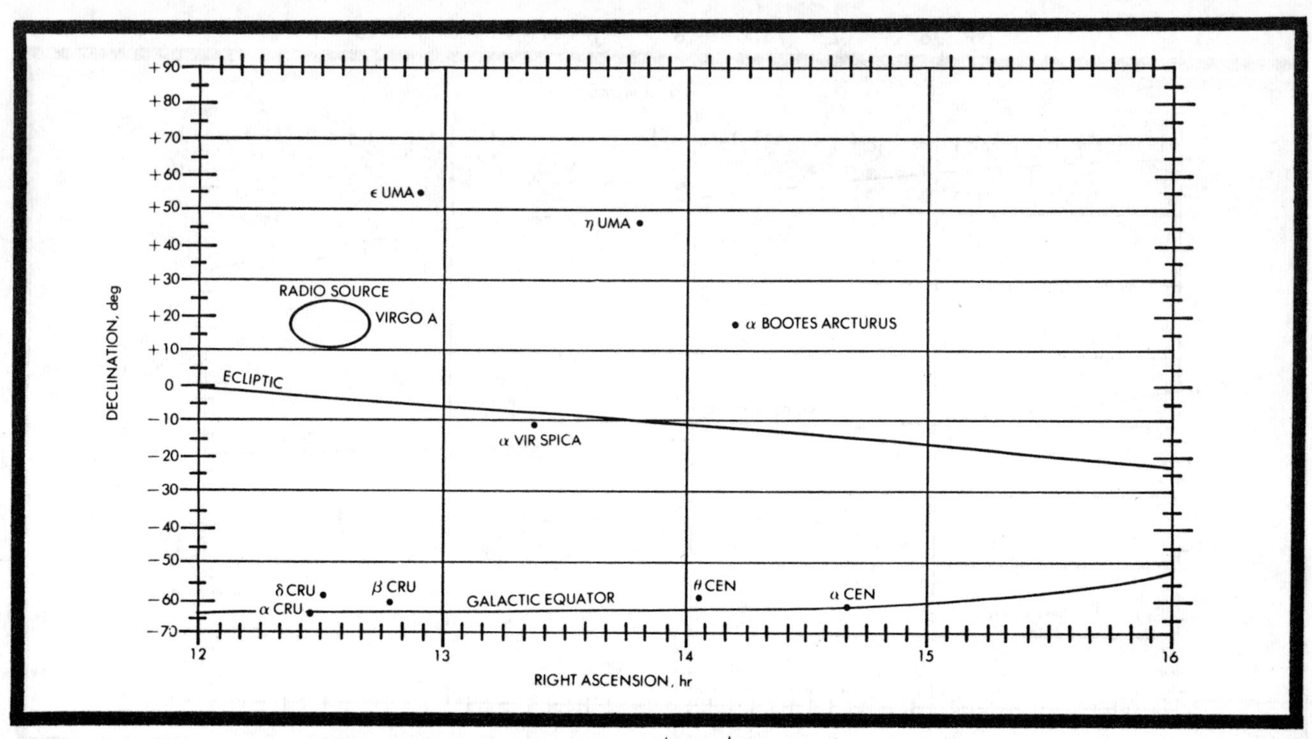

Fig. 4-9. A basic celestial map—12^h to 16^h RA, −70° to +90° dec.

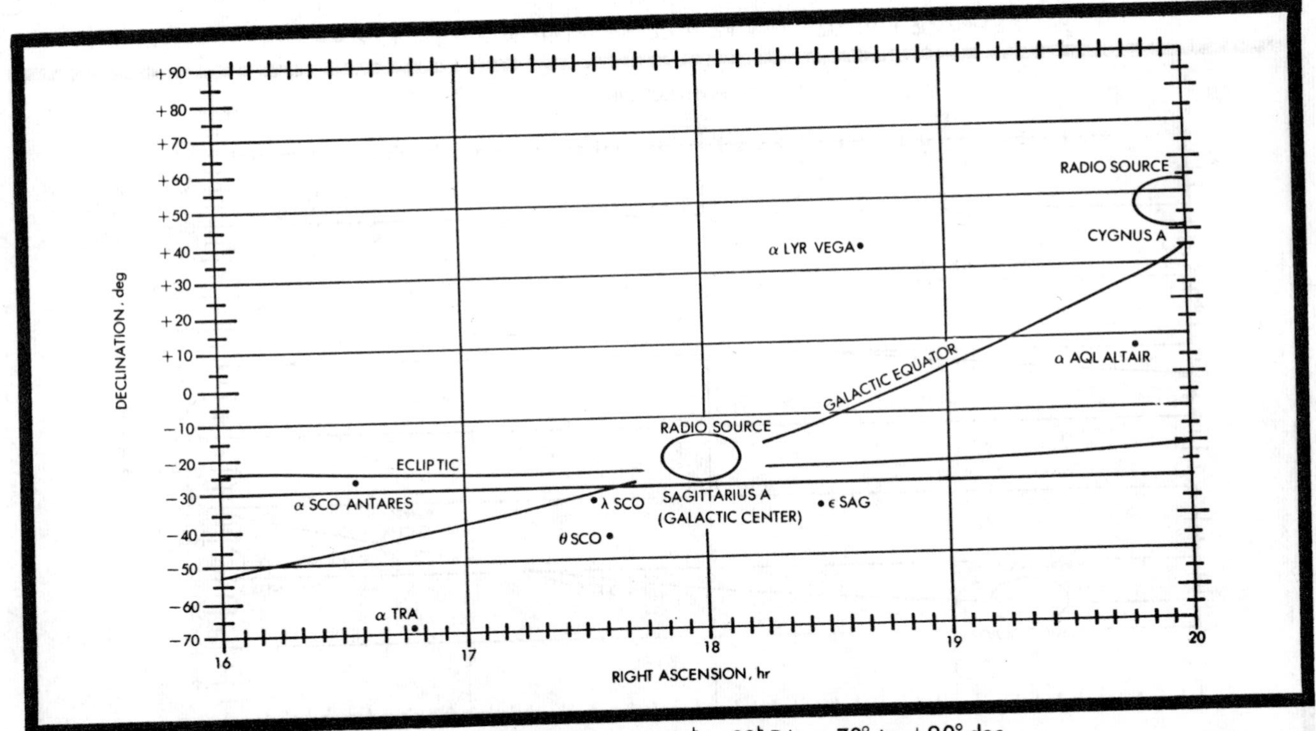

Fig. 4-10. A basic celestial map—16^h to 20^h RA, −70° to +90° dec.

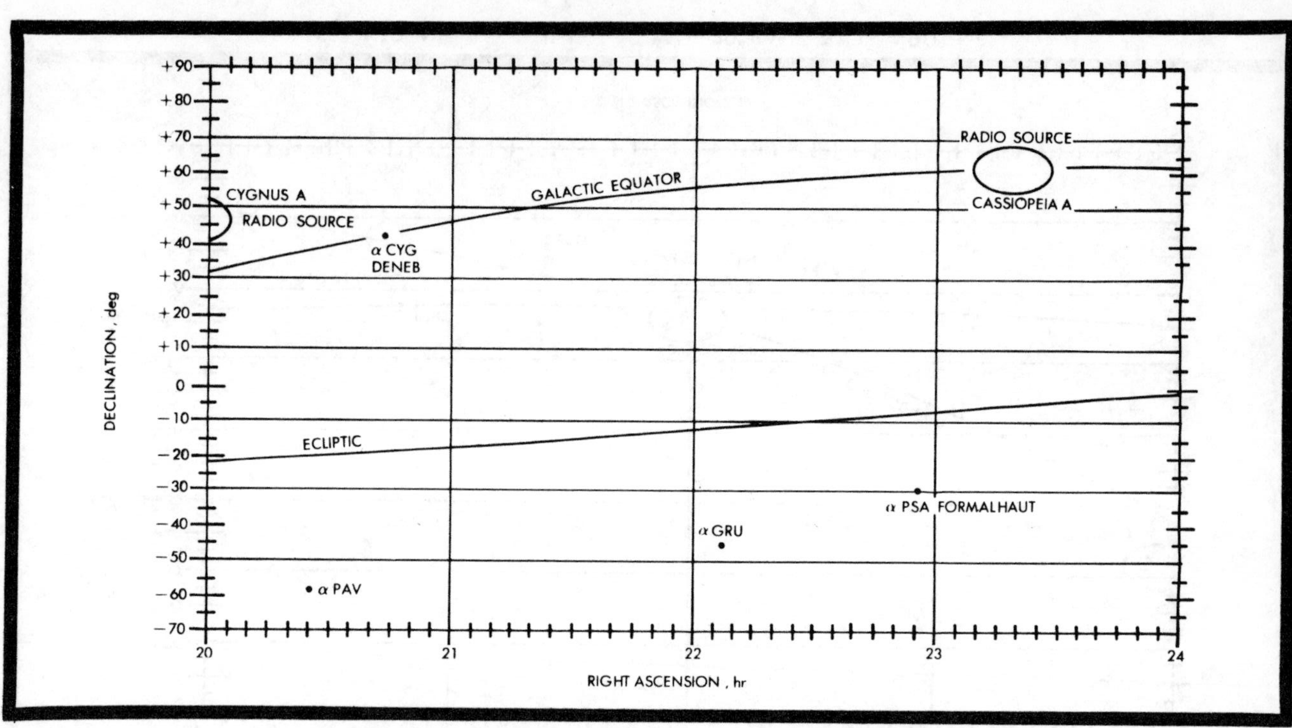

Fig. 4-11. A basic celestial map—20^h to 24^h RA, $-70°$ to $+90°$ dec.

Using the celestial maps the other way around, suppose an amateur finds an unlisted radio source at 5^h30^m RA and $-15°$ dec. The map in Fig. 4-7—the one covering the sky from 4^h to 8^h of right ascension—shows those particular coordinates to be a point in the sky just south of the constellation of *Orion*. Make up a number of arbitrary coordinates, and practice positioning them on the celestial maps.

USING CELESTIAL COORDINATES

The declination (dec) and right ascension (RA) of a star or radio source fix its position on the celestial sphere. The celestial sphere, however, is not fixed with respect to Earth. As a matter of convenience, astronomers assume the Earth is perfectly motionless and that the celestial sphere rotates about it. According to this point of view, the celestial sphere rotates on its polar axis once every $23^h56^m04^s$ of *mean solar* time. A sidereal day, in other words, is about 3^m56^s shorter than a 24-hour mean solar or *civil* day.

Given the celestial coordinates of a point, finding that point in the sky is a matter of determining the orientation of the celestial sphere at a certain time and from a certain place on the surface of the Earth. Much of the mathematical work in astronomy is devoted to determining the orientation of the celestial sphere and translating that information into positional coordinates that are most useful and meaningful to the observer.

The following sections describe some ways to determine the orientation of the celestial sphere as seen from any point on Earth at any time of the day on any day of the year.

HORIZON COORDINATES

Everyone has at least an intuitive notion of what a horizon might be—a line where the Earth meets the sky. But for astronomical purposes, the notion of a horizon must be defined in more precise terms. In astronomical work, the horizon is defined as a plane that runs through the center of Earth and is perpendicular to the observer's local pull of gravity. (See Fig. 4-12.) This particular definition of the horizon is not valid for civil and military navigation purposes because an aircraft

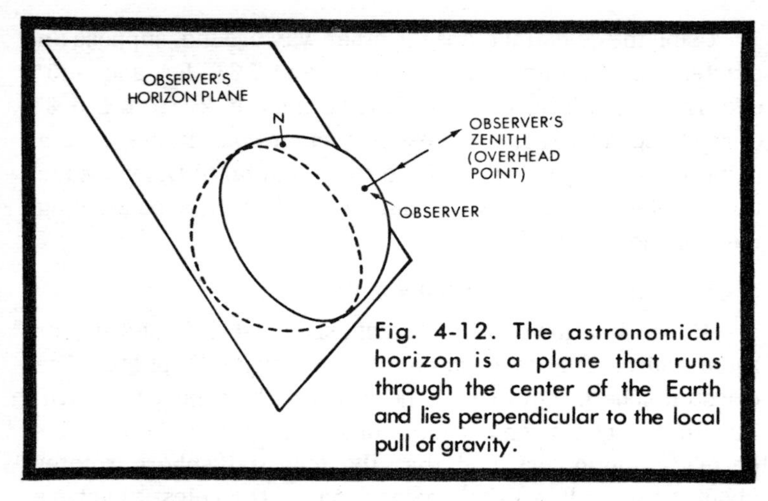

Fig. 4-12. The astronomical horizon is a plane that runs through the center of the Earth and lies perpendicular to the local pull of gravity.

flying over the Earth's equator, for example, is far below the horizon for an observer at 40°N latitude. The distance between the Earth and the nearest celestial bodies are so great, however, that bodies directly over the Earth's equator are visible from points close to the geographic north pole. In fact, an observer at 40°N latitude can see celestial bodies standing over 50°S latitude.

The local horizon, then, is determined in terms of the Earth's pull of gravity rather than the visible line between Earth and sky. An observer standing in a deep valley gets a false impression of the true horizon if he uses the sky–Earth interface as a reference rather than the local pull of gravity.

A simple way to determine the true horizon is by hanging a plumb bob (string and weight) from a yardstick. Hold the yardstick in a horizontal position and adjust it so the bob hangs at a right angle to it. With the apparatus thus adjusted, the bob indicates the direction of local gravity and the stick points directly toward the local horizon.

Most beginning amateur radio astronomers, especially those who have a background in military radar or microwave communications, are acquainted with the azimuth–altitude or *horizon* system of coordinates. It is the system commonly used to specify the coordinates of an object in the sky relative to the observer's horizon. And in this system, a point in the sky is specified in terms of two angles: one indicating the angular distance from true north and another indicating the angular distance from the horizon.

Azimuth is defined as the angular distance of a point in the sky as measured from true north through a 360° circle parallel to the horizon. In such a system, true north is taken as the 0° reference point, and all successive measurements run eastward from that point. An object due east of an observer, for example, has an azimuth angle of 90°. An object due south of the observer, on the other hand, has an azimuth of 180°.

Altitude is defined as the angular distance of a point in the sky as measured from the true horizon through a 90° arc to the overhead point, or *zenith*. The true horizon is taken as the 0° reference point, and all successive readings run upward to the zenith from that point. (See Fig. 4-13.)

A point in the sky that is due south of the observer and about halfway between the horizon and an overhead point has horizon coordinates of 180° az and 45° alt. As another example, a point having horizon coordinates of 270° az and 10° alt is just a bit above the western horizon.

Whereas a horizontal great circle defines the observer's astronomical horizon, a vertical great circle running from the observer's northern horizon, through the overhead point, the down to the southern horizon defines the observer's *meridian*. In terms of celestial coordinates, the meridian circle runs from 0° az 0° alt, through 90° alt, and to 180° az 0° alt.

SIDEREAL TIMEKEEPING

As indicated at the beginning of the previous section, the procedure for locating a given point in the sky is a two-step process. The observer must first determine the orientation of the celestial sphere as seen from the time and place of interest. He then must translate the celestial coordinates into terms that are meaningful to his view of the horizon.

Determining the orientation of the celestial sphere is a matter of determining the local sidereal time. The procedure for reckoning sidereal time might seem confusing at first, but it soon becomes second nature for anyone working with radio astronomy for any length of time.

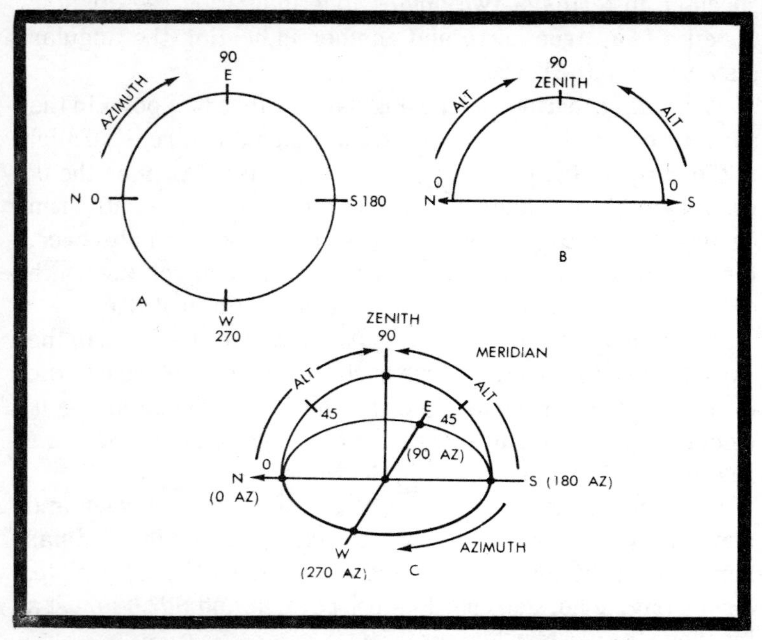

Fig. 4-13. The horizon system of coordinates. In A, azimuth is marked off along the astronomical horizon, beginning with 0° at north and running eastward through a full 360°. Altitude (B) is marked off along the observer's meridian—an imaginary semicircle running between the observer's southern horizon, through his overhead point (zenith), and down to his northern horizon. Altitude is measured from 0° at the horizon to 90° at the zenith.

Civil time standards are all based upon the motion of the Sun across the sky. Averaging figures for the Sun's time of rising and setting over an interval of one year, it spends about 12 hours per day in the sky and 12 hours below the horizon. The mean solar day is thus defined as exactly 24 hours. Whereas the Sun actually spends more time in the sky during the summer months than it does during the winter, the celestial sphere rotates at the same rate year in and year out. What is even more important is the fact that the celestial sphere rotates a bit faster than the Sun does on the average.

The celestial sphere rotates through a full 360° in $23^h56^m04^s$ of solar mean time. One complete rotation of the celestial sphere defines a *sidereal day*, and the astronomical timekeeping system based upon the sidereal day is called *sidereal time*.

A sidereal day is divided into 24 equal sidereal hours, each hour being approximately 9.8 seconds shorter than a civil hour. Every sidereal hour is likewise divided into 60 sidereal minutes and then into 60 sidereal seconds.

Mean solar time and sidereal time are both expressed in terms of hours, minutes, and seconds. Two minutes past midnight in mean solar time can be expressed as 0:02 a.m., 0002 hours, or $0^h2^m0^s$. Two minutes past midnight sidereal time is normally expressed in the form: $0^h2^m0^s$. These formating conventions bring up the possibility of confusion when using the *h-m-s* form; but unless it is perfectly clear which time system is being used, the time expression is designated as one version of mean solar time (such as UT or EDST) or sidereal time (ST).

Another important difference between civil and sidereal time is that the latter is different for every increment of longitude on the Earth. Sidereal time is based upon the position of the celestial sphere relative to an observer's longitude; and even if the longitudinal difference between the two observers is only a minute or two or arc, their sidereal times are correspondingly different. This scheme is quite different from civil time-keeping conventions that are often fixed by some arbitrary time-zone divisions.

Midnight in sidereal time is defined as the moment the First Point of Aries crosses the observer's meridian. And since the celestial sphere rotates faster than the mean Sun does, sidereal midnight occurs during the evening hours at one time of the year and during the daylight hours at other times.

The importance of sidereal time becomes clear when one realizes that sidereal time is equal to the right ascension of the point on the local meridian. If, for instance, a point on the meridian has a right ascension of 14^h23^m, the sidereal time for the observer is 14^h23^m. One of the easiest ways to locate a celestial object is to note its right ascension on a star table and keep track of the sidereal time. When the sidereal time is equal to the object's right ascension, the observer can be confident the object is right on his meridian.

Astronomers use sidereal clocks—clocks that run about 1.01 times faster than a normal clock—to keep track of the sidereal time. Since sidereal clocks have different gear ratios than their civil counterparts, and since there is relatively little demand for them, regular sidereal clocks can be rather expensive.

Amateur astronomers often use a normal clock to keep track of sidereal time, however. Once it is set to the correct sidereal time, the error over a 24-hour period is insignificant for most applications. Of course, there is the problem of keeping track of the difference between 0300 hours (3 a.m.) and 1500 hours (3 p.m.) on a normal 12-hour clock.

Another simple way to keep a running account of sidereal time is with the help of a battery-operated clock. Most battery-operated clocks have a speed adjustment that can be used to make the clock run about 3 minutes fast each day.

Whether an astronomer uses a sidereal clock or not, he must be able to determine his sidereal time by either noting the right ascension of a star on his meridian or by using a sidereal timetable. Discussions throughout this book assume the experimenter does not have access to a sidereal clock, and even those using such a clock have to apply the same procedures to set the clock from time to time.

The sidereal timetable in Appendix C shows the sidereal time at midnight in Greenwich for four-day intervals throughout the year. The following equation can be used to calculate the local sidereal time, given the sidereal time at midnight in Greenwich:

$$ST = ST_{GM} + LT + 0.0027\left(\frac{L}{15} + LT\right) \qquad (4\text{-}2)$$

where

ST = local sidereal time

ST_{GM} = sidereal time in Greenwich at midnight on the date in question

LT = local standard time

L = the observer's west longitude.

To see how this equation works, consider the following example: An observer located at 80°W longitude wants to know his sidereal time for 1500 hours EST on March 11.

From Appendix C, $ST_{GM} = 11.34^h$ on March 11. His local standard time is given in the problem as 15^h, and his longitude is 80°. Substituting this information:

$$
\begin{aligned}
ST &= 11.34+15+(0.0027)\ (80/15+15) \\
&= 26.34+(0.0027)\ (20.33) \\
&= 26.34+0.05 \\
&= 26.39 \text{ or } 2.39^h \text{ on the next day, March 12.}
\end{aligned}
$$

This particular example shows that the observer's sidereal time at 3 p.m. on March 11 is 2^h23^m. He thus knows he can calibrate his sidereal clock to 2:23 at 3 p.m. More importantly, he knows that celestial bodies with a right ascension of 2^h23^m will cross his meridian at 3 p.m. on March 11.

The sidereal timetable in Appendix C shows the values of ST_{GM} for about every fourth day of the year. Use the following equation to determine ST_{GM} for all other days of the year.

$$ST_{GM} = ST'_{GM}+0.066N \qquad (4\text{-}3)$$

where

ST'_{GM} = the last known Greenwich midnight sidereal time

N = the number of days since the given ST'_{GM}.

If, for example, an experimenter wants to know the Greenwich-midnight sidereal time on May 8, he can use the table in Appendix C to determine the sidereal time on the nearest previous date and apply the equation. By Appendix C, the value of ST'_{GM} is 15.02^h on May 6. The number of days to May 8, the value of N, is equal to 2. Thus:

$$ST_{GM} = 15.02+(0.066)\ (2) = 15.15^h \text{ or } 15^h9^m.$$

The equation can be used to determine ST_{GM} for any day of the year without the help of a sidereal timetable. The sidereal time at midnight in Greenwich on March 20 is always very close to 12^h. By figuring the number of days from March 20 to the date of interest, an experimenter can substitute 12^h for ST'_{GM} and the number of days from March 20 for N.

Suppose the experimenter wants to know the value of ST_{GM} on September 1. According to the "Days Between Two Dates" table in the *World Almanac*, the number of days between March 20 and September 1 is 165.

Then: $ST_{GM} = 12+0.066(165) = 22.89^h$ or 22^h53^m.

The constant multiplier of 0.0027 that appears in Eq. 4-2 can be rather troublesome for experimenters who do not have access to an electronic calculator. It is possible to rough in an estimate of the local sidereal time by simply ignoring the terms multiplied by 0.0027, and considering sidereal time to be roughly equal to ST_{GM} plus LT. The example following Eq. 4-2 shows a calculated local sidereal time of 2.39^h. Using the estimating technique, the local sidereal time becomes $11.34+15 = 26.34^h$ or 2.34^h (after subtracting a 24-hour day). The rough estimate is only 0.05 hour (3 minutes) low compared to the more accurate result. The difference is so small that most experimenters prefer to use the rough estimate for everyday calculations, leaving the more accurate procedure for clock calibration and precise optical work.

HORIZON–CELESTIAL CONVERSIONS

The procedures for locating, identifying, and specifying any optical or radio source is closely associated with the specification of the source's position in celestial coordinates. And since the actual position of an object on the visible celestial sphere depends upon the observer's position on Earth and the time of day, some computations are unavoidable.

The computations can be greatly simplified if the experimenter happens to use an equatorial antenna mount as opposed to an altiazimuth version. An equatorial mount, properly scaled in celestial coordinates, completely eliminates the need for horizon-to-celestial conversions, leaving only the computations needed to determine the sidereal time.

Few amateur radio astronomers care to construct an elaborate equatorial antenna mount, so they must learn to live with something of a mathematical burden. As demonstrated in this section, however, it is possible to minimize the computations by restricting observations to objects on the local meridian. The more general equations also appear in this section for the benefit of those who do not wish to wait for a particular source to cross the meridian.

Meridian Transit Computations

Every celestial object visible from Earth crosses or *transits* the meridian at least one time each day. Objects that never drop below the horizon—those that rotate rather close to Polaris—make two meridian transits each day: one on the southern leg of the meridian (an *upper culmination*) and another on the northern leg of the meridian (a *lower culmination*). Most celestial bodies only undergo the lower culmination each day.

Amateur radio astronomers routinely face two kinds of computational situations:

1. Given the celestial coordinates of a known source, determine the time of transit and altitude on the meridian.
2. Given the time of transit and altitude of an unknown object on the meridian, determine its celestial coordinates.

The first computational situation is essentially a prediction problem. The observer knows what source he wants to observe, and the problem is to determine exactly when and where it will cross his meridian. The second situation is applied when the experimenter is attempting to identify a source he observes on his equipment. By determining the celestial coordinates of the source in the second case, he can correlate the position with a known source on a celestial map.

Equations 4-4 through 4-6 used in this chapter apply the following notations:

LT = local standard time (subtract 1^h for DST)
L = the observer's geographical longitude (°W)
ST_{GM} = the sidereal time in Greenwich at 0000^hUT
RA = the right ascension of the source
dec = the declination of the source
alt = altitude of the source on the meridian
lat = the observer's geographical latitude (°N)

Determining time of transit and altitude is a matter of solving Eqs. 4-4a and 4-4b.

$$LT = 0.99(RA-ST_{GM})-0.0002L \quad (4\text{-}4a)$$
$$alt = dec-lat+90° \quad (4\text{-}4b)$$

Equation 4-4a is used whenever an experimenter wants to know *when* to look for a certain star or source on his local meridian. All he has to know is the right ascension of the source, the ST_{GM} (Appendix C) for the day he wants to look for the source, and his own geographical longitude. The following example illustrates the use of Eq. 4-4a.

Example 4: What is the time of transit of Cygnus A on June 11, if the observer's position is 80°W? Solving 4-4a requires a knowledge of the right ascension of Cygnus A and the ST_{GM} for June 11—Appendix B provides the right ascension information and Appendix C gives the sidereal time data. From Appendix B, the right ascension of Cygnus A is about 20^h. The ST_{GM} for June 11 is 17.39^h. Solving Eq. 4-4A with this information:

$$\begin{aligned} LT &= 0.99(20-17.39)-0.0002(80) \\ &= 0.99(2.61)-0.016 \\ &= 2.58-0.016 \\ LT &= 2.56^h \text{ or } 2{:}34 \text{ a.m. EST.} \end{aligned}$$

The experimenter thus knows he can expect to find Cygnus A crossing his meridian at 2:34 a.m. on June 11.

Equation 4-4b is used whenever an experimenter wants to know how far above the horizon the source will be when it does cross the meridian. The result of the foregoing example shows when the transit will occur on June 11, but it doesn't tell the experimenter where to look for it. Solving Eq. 4-4b does the job.

Example 5: The experimenter in the previous example is located at latitude 40°N. Determine the altitude of Cygnus A when it transits on June 11.

From Eq. 4-4b:

$$alt = 59-40+90 = 109°$$

The source, in other words, will be 109° above the southern horizon. It is perhaps more meaningful to express the meridian position with reference to the northern horizon, however. To do this, subtract the altitude from 180°: alt = 180 − 109 = 71° over the northern horizon.

Summarizing the results of the problem in Examples 4 and 5, the experimenter can expect to find Cygnus A 71° above the northern horizon at 2:34 a.m. on June 11.

Equation 4-4a, for determining the local time of transit of a source, contains two constant multipliers that can be rather troublesome as far as quick-and-easy solutions are concerned. It is possible to round off Eq. 4-4a so that it provides relatively reliable results without your having to go through some of the tricky multiplication with constants. For many practical purposes, it is possible to restate Eq. 4-4a as:

$$LT = RA - ST_{GM}$$

The simplified version of Eq. 4-4a can be solved in a few seconds; and with some experience, an astronomer can learn to make the calculations in his head. The accuracy isn't bad, either. Try the problem in Example 4. In that problem, $RA = 20$ and $ST_{GM} = 17.39$. According to the simplified version of the basic equation:

$$LT = 20 - 17.39 = 2.61^{h} \text{ or } 2{:}37 \text{ a.m.}$$

Compared to the accurate result of 2:34 a.m., the result of the simplified operation is only 3 minutes low; and a 3-minute difference cannot be detected by most amateur radio telescopes.

Determining the celestial coordinates of an unknown source is a matter of rearranging the terms in Eqs. 4-4a and 4-4b.

$$RA = ST_{GM} + 1.0027LT + 0.0002L \qquad (4\text{-}5a)$$

$$dec = lat + alt - 90° \qquad (4\text{-}5b)$$

Equation 4-5a is used whenever an experimenter wants to determine the right ascension of an unknown source that is on his meridian. Equation 4-5b lets him determine the declination of that same source. The two equations thus enable the experimenter to completely determine the celestial coordinates of any object on his meridian, provided he has access to his local standard time, geographical latitude and longitude, and the sidereal time in Greenwich at 0^h UT (Appendix C).

Example 6: What is the right ascension and declination of a source observed at 11 a.m. EST on July 5, if the antenna was set on the meridian at an altitude of 72°? Assume the observer is located at 40°N, 80°W.

According to Eq. 4-5a:

$$\begin{aligned} RA &= 18.96+1.0027\,(11)+0.0002\,(80) \\ &= 18.96+11.03+0.016 \\ RA &= 30^h \end{aligned}$$

Subtract 24^h to tidy up the result: $30 - 24 = 6^h$.

By Eq. 4-5b, the declination of the source is:

$$dec = 40+72 - 90 = +22°.$$

The celestial coordinates of the observed source are thus 6^h right ascension and +22° declination. A survey of a celestial star map or list of known radio sources (Appendix B) shows that *Taurus A* has coordinates $5^h31^m30^s$ at a declination of $21°50^m$. There is little room for doubt that the observed source is closely associated with Taurus A. Similar results observed on several consecutive days could confirm this conclusion.

The formula for finding the right ascension of a source on the meridian, Eq. 4-5a, can be somewhat simplified for most practical purposes. Rounding off the constant multipliers in that equation produces:

$$RA = ST_{GM}+LT.$$

Applying the information given in Example 6, the estimated right ascension of the source is equal to 18.96+11 or 29.96^h. After subtracting 24^h, the estimated RA is 5.96^h—0.04^h less than the accurate figure determined after multiplying the figures by some tricky constants. The error is tolerable; it is, in fact, lost in the cumulative Pythagorean error of the measuring instruments.

Determining convenient times of observation is a problem that is especially relevant for amateur astronomers. Except for weekends and holidays, few amateur radio astronomers can make direct observations of meridian crossings at just any old time of the day. And in some locations, prevailing radio interference makes daylight observations virtually impossible during normal working hours.

It is thus helpful to calculate the dates a given radio source will cross the meridian at hours most convenient for the experimenter. Stated formally: Given the set of times for making calculations, determine the nearest dates a certain object makes its meridian transits.

Rearranging Eq. 4-5a to solve for the sidereal time of meridian crossing produces:

$$ST_{GM} = RA - 1.0027LT - 0.0002L \qquad (4\text{-}6)$$

Like the other equations in this section that are related to right ascension, Eq. 4-6 can be rounded off to provide reasonably accurate answers for a smaller amount of mathematical work. In this instance, the equation becomes:

$$ST_{GM} = RA - LT$$

Equation 4-6 and its estimated version do not directly give the experimenter the observation dates, but they do provide an ST_{GM} which can, in turn, be used to determine the dates from Appendix C.

Example 7: Suppose an experimenter has a work schedule that allows him to make direct radio observations of the meridian between 10 p.m. and midnight EST. Between what two dates will he be able to observe *Cassiopeia A* on the meridian? Assume his location is 40°N, 83°W.

Working the simplified version of Eq. 4-6 requires a knowledge of the source's RA and the time of day that the observer can be on the job. From Appendix B, the RA of Cassiopeia A is 23^h21^m, or 23.35^h. The observer's earliest working time is 10 p.m., or 22^h. Plugging these values into the estimated form of Eq. 4-6:

$$ST_{GM} = 23.35 - 22 = 1.35^{h.}$$

This experimenter can thus expect to observe Cassiopeia A on the meridian at 10 p.m. on the night when the sidereal time at Greenwich midnight is 1.35^h. The sidereal timetable in Appendix C shows this occurs on October 9.

Repeating the procedure for the midnight (0^h) observing time:

$$ST_{GM} = 23.35 - 0 = 23.35^h.$$

Cassiopeia A thus appears on the observer's meridian at midnight EST when the sidereal time at Greenwich midnight is 23.35^h. The sidereal timetable shows this happens around September 11.

Putting together the results of both sets of computations, the observer can expect to find Cassiopeia A on the meridian between 10 p.m. and midnight EST from September 11 through October 9. He can determine the altitude of that source using Eq. 4-4b.

Computations for Off-Meridian Sources

The computations required for converting back and forth between the *celestial* and *horizon* coordinate systems become quite complicated for sources not on the local meridian. The formulas are so awkward, in fact, that they are normally omitted from basic astronomy references. To get around the need for such operations, amateurs usually have the choice of either using an equatorial mount or working only with meridian-transit phenomena.

Amateur radio astronomers seldom want to build a cumbersome equatorial mount for their larger antennas, and they often feel their work is severely limited when observing only meridian transits. The off-meridian equations described in this section are, indeed, rather complex; but the current proliferation of low-cost electronic calculators makes the routine solution of such problems easier than ever before.

The two basic coordinate transformation situations are quite similar to those required for meridian transit studies. The off-meridian equations are more general, however; and the simpler meridian equations are actually derived from them. The situations calling for the mathematical transformations are:

1. Given the celestial coordinates of a known source and a time of observation, determine the source's azimuth and altitude.

2. Given the time an unknown source appears at a certain azimuth and altitude, determine its celestial coordinates.

The mathematical notations described for meridian-transit calculations in the previous section hold for the following off-meridian equations as well. The off-meridian calculations introduce two more variables, however:

az = azimuth of the source in degrees
HA = hour angle of the source in degrees

Determining the hour angle of a source. The hour angle of a source is the angular distance between the observer's meridian and the source as measured in a westerly direction along the celestial equator. (See Fig. 4-14.)

Equations 4-7a and 4-7b show the simple relationship between the local sidereal time, the right ascension of an object, and its hour angle.

$$HA = 15\,(ST - RA) \qquad (4\text{-}7a)$$

$$RA = ST - \frac{HA}{15} \qquad (4\text{-}7b)$$

where ST = the local sidereal time from Eq. 4-2

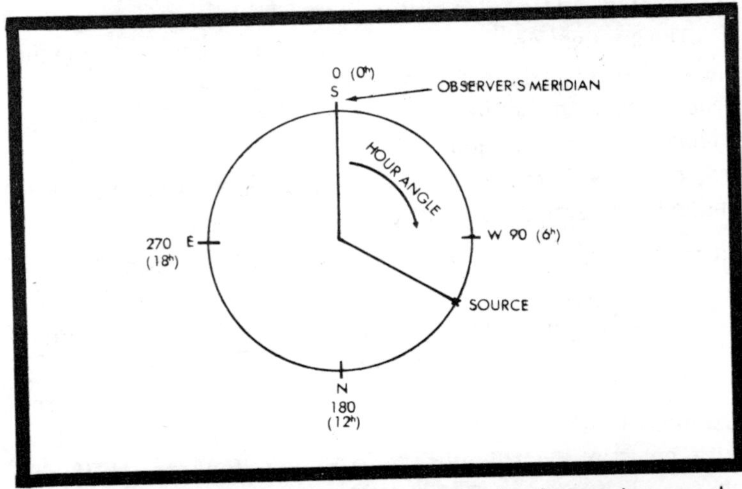

Fig. 4-14. The hour angle of a source is the angular distance between the local meridian and a source on the celestial sphere. Hour angles are measured from south (the meridian) and through a full 360° in the westerly direction.

Equation 4-7a is used to determine the hour angle of an object in degrees when the local sidereal time and right ascension of the object are known. Equation 4-7b has been rearranged to accommodate the situations where it is necessary to solve for the right ascension of an object from its hour angle and the sidereal time.

Example 8: Suppose an observer wants to know the hour angle of a source having a right ascension of 23^h40^m when the sidereal time, from Eq. 4-2, is found to be 13^h.

From Eq. 4-7a: $HA = 15(13-23.66) = -159.9$

$$HA = 360-159.9 = 200.1^\circ$$

The first result in this example, -159.9°, has a negative value that indicates the hour angle is 159.9° east of the observer's meridian. Hour angles are normally expressed in positive degrees west, however, so the figure is adjusted accordingly. The final result, $\approx 200^\circ$, indicates the source is that far west of the observer's meridian.

Determining Altitude and Azimuth of Known Sources. Equations 4-8a and 4-8b are used for converting the known celestial coordinates of a point in the sky into horizon coordinates expressed in altitude and azimuth.

The application of these equations requires a good working knowledge of trigonometry. The experimenter must be acquainted with the application of related angles and inverse trigonometric functions in particular. An ordinary trigonometry textbook can be an invaluable aid to those whose understanding might be a little rusty.

$$\text{alt} = \sin^{-1}[\sin(\text{dec})\sin(\text{lat})+\cos(\text{HA})\cos(\text{dec})\cos(\text{lat})] \quad (4\text{-}8a)$$

$$\text{az} = \cos^{-1}\left[\frac{-\cos(\text{HA})\cos(\text{dec})\sin(\text{lat})+\sin(\text{dec})\cos(\text{lat})}{\cos(\text{alt})}\right] \quad (4\text{-}8b)$$

Example 9: What will be the altitude and azimuth of a source with celestial coordinates 2^h, $+59^\circ$ on January 11 if the observer is located at 40°N, 80°W?

The solution of Eqs. 4-8a and 4-8b requires a knowledge of the hour angle which, in turn, uses the information about the date, local time, and longitude. From the sidereal timetable in

Appendix C, the ST_{GM} for January 11 is about 7.58. The estimated version of Eq. 4-2 is $ST = ST_{GM}+LT$. Thus, the local sidereal time at 3:30 p.m. is:

$$ST = 7.58+15.5 = 23.08^h.$$

The hour angle of the source, from Eq. 4-7, is:

$$HA = 15(23.08 - 2) \text{ or about } 316°.$$

Rounding off the appropriate information from standard trigonometric tables or the simplified tables in Appendix D:

$$\sin(\text{dec}) = \sin 59° = 0.86 \qquad \sin(\text{lat}) = \sin 40° = 0.64$$
$$\cos(\text{dec}) = \cos 59° = 0.52 \qquad \cos(\text{lat}) = \cos 40° = 0.77$$
$$\cos(\text{HA}) = \cos 316° = 0.72$$

From Eq. 4-8a:

$$\begin{aligned} \text{alt} &= \sin^{-1}[(0.86)(0.64)+(0.72)(0.52)(0.77)] \\ &= \sin^{-1} 0.84 \\ &\approx 57° \end{aligned}$$

The only additional information required for finding the azimuth is the cosine of the altitude: $\cos 54° = 0.59$.

$$\begin{aligned} \text{az} &= \cos^{-1}\left[\frac{-(.72)(.52)(.64)+(.86)(.77)}{0.59}\right] \\ &= \cos^{-1} 0.72 \\ &= 44° \end{aligned}$$

At 3:30 p.m. EST on January 15, the observer can expect to find a source having celestial coordinates of 2^h, $+59°$ at a position 44°az, 57°alt with respect to his view of the horizon.

Figure 4-15 shows a worksheet that is especially helpful for solving Eqs. 4-8a and 4-8b with the aid of a calculator that has no memory function.

Determining Right Ascension and Declination of a Source. Much of the work in radio astronomy concerns translating the horizon coordinates of a position into celestial coordinates. The objective in this instance is to fix the object's position on a regular celestial map and correlate it with some observable object in the sky.

Equations 4-19a and 4-19b are used to convert the horizon coordinates of a point in the sky into the corresponding

ALTITUDE AND AZIMUTH OF OFF-MERIDIAN SOURCES

Preliminary Data: declination of source = ________

sin (dec) = ______

cos (dec) = ______

local latitude = ___ ___ W

sin (lat) = ______

cos (lat) = ______

hour angle of source = ________

cos (HA) = ______

ALTITUDE

sin (dec) × sin (lat) = ______

cos (HA) × cos (dec) × cos (lat) = ______

sum = ______

Sin $^{-1}$ (sum) =

Altitude of the source

cos (alt) = ______

(for azimuth computations)

AZIMUTH

sin (dec) × cos (lat) = ______

cos (HA) × cos (dec) × sin (lat) = ______

sum = ______

− cos (alt) = ______

cos $^{-1}$(quotient) = ______

azimuth of the source

ALTITUDE AND AZIMUTH OF OFF-MERIDIAN SOURCES

Preliminary Data: declination of source = 59

sin (dec) = .857

cos (dec) = .515

local latitude = 40

sin (lat) = .643

cos (lat) = .766

hour angle of source = 100

cos (HA) = −.174

ALTITUDE

sin (dec) × sin (lat) = .551

cos (HA) × cos (dec) × cos (lat) = −.069

sum = .482

Sin $^{-1}$ (sum) = 29

Altitude of the source

cos (alt) = .875

(for azimuth computations)

AZIMUTH

sin (dec) × cos (lat) = .656

cos (HA) × cos (dec) × sin (lat) = −.058

sum = .714

− cos (alt) = .816

cos $^{-1}$ (quotient) = 35

azimuth of the source

29 alt

35 az

Fig. 4-15. Worksheets (and worked-out example) for determining the altitude and azimuth of an off-meridian source.

declination and hour angle. The hour-angle term can be converted into right ascension by means of Eq. 4-7b.

$$\text{dec} = \sin^{-1}[\cos(\text{az})\cos(\text{alt})\cos(\text{lat})+\sin(\text{alt})\sin(\text{lat})] \quad (4\text{-}9a)$$

$$\text{HA} = \cos^{-1}\left[\frac{\sin(\text{alt})\cos(\text{lat}) - \cos(\text{az})\cos(\text{alt})\sin(\text{lat}}{\cos(\text{dec})}\right] \quad (4\text{-}9b)$$

Example 10: Suppose an experimenter finds an interesting radio response when his antenna is set at azimuth 220° and altitude 30°. The time of the observation is 1:30 a.m. EST on November 6. What are the right ascension and declination of the source if the observer's position on the earth is 40°N, 80°W?

Rounding off the appropriate information from standard trigonometric tables or the simplified tables in Appendix D:

$\sin(\text{alt}) = \sin 30° = 0.5$ $\qquad \sin(\text{lat}) = \sin 40° = 0.64$

$\cos(\text{alt}) = \cos 30° = 0.87$ $\qquad \cos(\text{lat}) = \cos 40° = 0.77$

$\cos(\text{az}) = \cos 220° = -\cos 40° = -0.77$

Applying Eq. 4-9a:

$$\begin{aligned}\text{dec} &= \sin^{-1}[(-0.77)(0.87)(0.77)+(0.5)(0.64)]\\ &= \sin^{-1} -0.20\\ &\approx -12°\end{aligned}$$

The equation for hour angle cannot be applied until cos(dec) is found. From trigonometric tables, the value is: $\cos(-12°) = \cos 12° = 0.98$. Applying Eq. 4-9b:

$$\begin{aligned}\text{HA} &= \cos^{-1} \frac{(0.5)(0.77)-(0.77)(0.87)(0.64)}{0.98}\\ &= \cos^{-1} 0.81\\ &\approx 36°\end{aligned}$$

The hour angle can be converted into right ascension by means of Eq. 4-7b; but that equation requires a knowledge of the local sidereal time at 1:30 a.m. on November 6.

From the simplified version of Eq. 4-2 and the sidereal timetable in Appendix C:

$$\text{ST} = 3.11+1.5 = 4.61^{h}.$$

RIGHT ASCENSION AND DECLINATION OF OFF-MERIDIAN SOURCES

Preliminary Data: azimuth of the source = ___

cos (az) = ___

altitude of the source = ___

sin (alt) = ___
cos (alt) = ___

local latitude = ___ W

sin (lat) = ___
cos (lat) = ___

DECLINATION

cos (az) × cos (alt) × cos (lat) = ___
sin (alt) × sin (lat) = ___
sum = ___
Sin ⁻¹ (sum) = ___
dec of source
cos (dec) = ___
(for RA computations)

RIGHT ASCENSION

sin (alt) × cos (lat) = ___
cos (az) × cos(alt) × sin (lat) = ___
sum = ___
− cos (dec) = ___
cos ⁻¹ (quotient) = ___
HA of source
sidereal time = ___ hrs
minus HA/15 = ___
subtract = ___ hrs
right ascension of the source

RIGHT ASCENSION AND DECLINATION OF OFF-MERIDIAN SOURCES

Preliminary Data: azimuth of the source = 170

cos (az) = −.985

altitude of the source = 40

sin (alt) = .643
cos (alt) = .766

local latitude = 40 W

sin (lat) = .643
cos (lat) = .766

DECLINATION

cos (az) × cos (lat) × cos (lat) = −.577
sin (alt) × sin (lat) = .413
sum = −.164
Sin ⁻¹ (sum) = −9
dec of source
cos (dec) = .988
(for RA computations)

RIGHT ASCENSION

sin (alt) × cos (lat) = .493
cos (az) × cos (lat) × sin (lat) = + .485
sum = .978
cos (dec) = .990
cos ⁻¹ (quotient) = 8
HA of source
sidereal time = 21.5 hrs
HA 15 = .533
subtract = 20.967 hrs = $20^h 58^m$
right ascension of the source

RA = $20^h 58^m$
dec. = −9°

Fig. 4-16. Worksheets (and worked-out example) for determining the right ascension and declination of an off-meridian source.

Applying the local sidereal time of 4.61^h to Eq. 4-7b:

$$RA = 4.61 - 36/15 = 2.21^h \text{ or } 2^h\,13^m.$$

The source discovered in this example thus has a right ascenion of $2^h 13^h$ and a declination of $-10°$.

Figure 4-16 shows some sample worksheets for simplifying the mathematical procedures.

5 The Theory and Technology of Radio Astronomy

Radio astronomy is one of those recently discovered scientific disciplines that could have blossomed much sooner than it actually did. The physical principles behind the origin of some types of extraterrestrial radiation have been known from the beginning of this century, but no one thought about applying them to the stars. As fate would have it, the discovery of natural radio emissions from space had to be an accident.

This chapter describes both the theory of the origin of radio signals from space and the basic theory of radio telescopes from a technical point of view.

THE ORIGINS OF RADIO SIGNALS

Radio astronomers presently classify extraterrestrial signals as coming from thermal, nonthermal, and 21 cm hydrogen reactions. Some sources include the mechanisms for generating all three classes of signals, while others generate only one or two types. In any case, the signals are all electromagnetic in nature, and it is only their differences in quality that offer clues as to their origin.

Thermal Radiation

Any body that is heated above absolute zero generates broadband electromagnetic energy. The intensity of the emission and the spectral content are roughly proportional to the amount of heat; and in theory at least, it is possible to detect electromagnetic energy from any object in the universe.

To get an intuitive idea of the principle of thermal radiation, consider a familiar situation where heat is being applied to a bar of metal. Since the metal has a temperature above absolute zero from the start, it will be generating some broadband energy anyway. But consider what happens as the temperature rises. At some point in the heating process, the temperature reaches a point where the largest share of the radiation shifts to the visible part of the electromagnetic spectrum. The bar, in other words, starts to glow. As the temperature rises further, the spectral content shifts toward even shorter wavelengths, making the glow change from a cherry red toward a blue. A white-hot metal is one that is generating a great deal of broadband or *white* noise in the visible part of the spectrum.

The important point in this example is that a heated body does indeed emit electromagnetic energy. The problem is to detect that energy. Detecting the signals when most of the energy is in the visible part of the spectrum poses no real difficulty—the human eye acts as the receiver. Picking up the signals on either side of the visible part of the spectrum poses the main problem.

Radio receivers tuned below the visible part of the electromagnetic spectrum can detect the signals from a heated body, providing they are sensitive enough to do the job. And that's where a radio telescope comes into the picture.

Visible stars radiate a great deal of electromagnetic energy. Much of that energy has to be in the visible part of the spectrum—they would not be *visible* stars otherwise. Part of the energy has to be in the microwave portion of the spectrum, too; and that is the "fair game" professional radio astronomers seek out.

Not all visible stars are good radio transmitters, however. It turns out the hottest and brightest stars emit more energy at frequencies above the visible range than below. Such stars are noted for their X-ray and atomic-particle radiation.

Amateur radio astronomers can go after very intense thermal generators such as our own Sun and the Milky Way.

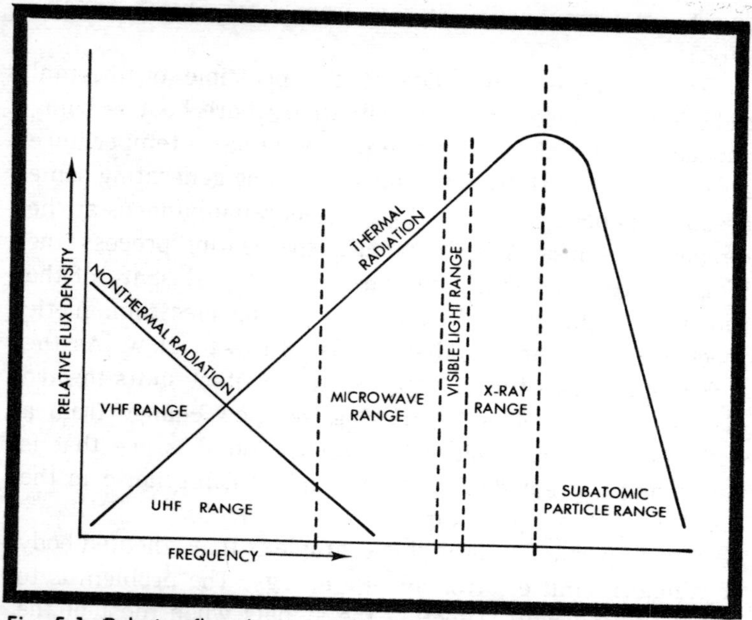

Fig. 5-1. Relative flux density of thermal and nonthermal sources as a function of frequency.

Radio signals of thermal origin have two characteristics that help distinguish them from other types of signals. Thermal radiation reproduces on a loudspeaker as pure static of the hissing variety. In fact, one of the main problems working with thermal radiation is that it is indistinguishable from the hissing produced by the radio circuitry itself.

A second feature of thermal radiation from extraterrestrial sources is that its intensity increases with frequency. Thermal radiation signals are practically nonexistent at the lower portion of the VHF band, but it can be present in immense doses at the upper UHF and gigahertz ranges. (See Fig. 5-1.)

Nonthermal Radiation

Much of the radiation from our own galaxy and most of that from other galaxies is of nonthermal origin. The mechanisms of nonthermal radiation are more complex and, in some instances, poorly understood compared to thermal mechanisms. Nonthermal radiation is a fortunate phenomenon for radio astronomy, however, because it enables radio telescopes to detect *cold* sources that would otherwise pass unnoticed.

Two theories of the origin of nonthermal radio energy from space prevail today. It is important to note at the outset that the theories do not compete with one another—they each explain the origin of two types of nonthermal radiation.

Whenever some mechanism such as high heat, intense magnetic or gravitational fields, or atomic radiation separate the electrons of an atom from their nucleus, those electrons take every opportunity to oppose those mechanisms and recombine with a positively charged ion. As the electron and ion accelerate toward one another, the electron emits electromagnetic energy. The external mechanisms tend to separate the electron and positive ion again at some later time, however, making the process continue indefinitely.

Whenever a vast quantity of free and oppositely charged ions coexist in a relatively small space, the net radiation from the reactions can add up to an intense, continuous wideband rf signal. Such conditions prevail around many stars in the galaxy and especially around nebulas and clusters of stars.

This particular phenomenon is known as *plasma oscillation*. Radio emissions from plasma oscillation never reach optical wavelengths under natural circumstances, and the signals are generally most intense in the lower VHF range. Plasma oscillation is a close relative to the mechanisms behind the laser principle, which is, of course, a purely artificial condition.

Synchrotron radiation, the second major class of nonthermal emission, takes its name from the fact that it was first discovered in connection with synchrocyclotron particle accelerators. Any moving electron gives off some radio energy; and when it is accelerated in a strong magnetic field to velocities approaching that of light, the resulting radio emissions can be quite strong.

Synchrotron radiation from space is characterized by its circular or elliptical polarization and its tendency to drift in amplitude. Most of the radiation from Jupiter is of the nonthermal, synchrotron variety, and a loudspeaker reproduces the emissions as whishing seashore sounds and bursts of clicks.

Like radio emission from plasma oscillation, the energy from synchrotron activity tends to taper off in the UHF bands. The fact that synchrotron radiation is unstable distinguishes it from plasma oscillation radiation; and the fact that both types of nonthermal radiation are most intense in the VHF range separates them from thermal radiation.

The third class of radiation of special interest to radio astronomers is 21 cm hydrogen-line radiation. All atoms, including those of hydrogen, radiate electromagnetic energy at a wavelength and intensity peculiar to the atom's composition. It is possible to identify the atoms of any element by noting the sharp, almost infinitely narrow radio-emission bandwidths.

It so happens that hydrogen atoms radiate this *resonance* energy at a constant frequency very close to 1428 MHz (1.428 GHz). Since hydrogen is the key substance in the universe, radio astronomers have a heyday locating thousands of radio sources by tuning their equipment to 1428 MHz. Unfortunately for amateurs, this 21 cm equipment is either too elaborate or too expensive for the average household budget. Until microwave equipment becomes available at a much lower cost, there is little point in pursuing the idea any further.

THE FUNDAMENTAL EQUATION

Physicists and electrical engineers tend to take slightly different views of radio astronomy. To a physicist, a radio telescope is one particular type of *radiometer*—a special divice used to determine the brightness temperature of a heated body. A physicist thus thinks about electromagnetic energy from space in terms of thermodynamics and *black body* radiation; and if the physicists had their way, radio astronomers would speak of signal intensities in terms of temperature—namely, degrees Kelvin.

Radio astronomy did indeed start off favoring the terminology of heat physics. The terminology created some problems for engineers designing the antennas and receiving systems, however. Instead of thinking in familiar units such as watts, meters, and hertz, the engineers were forced to adapt to temperature terms.

There are some advantages to using the temperature scheme for measuring signal strengths. Equation 5-1 shows the fundamental equation of radio astronomy as it is expressed in terms of brightness temperature.

$$S = \frac{kT_a}{A_e} \tag{5-1}$$

where

S = brightness flux density of the source in °K/m^2

T_a = the apparent brightness temperature of the antenna in °K

k = Boltzmann's constant, 1.38×10^{-27}

A_e = the effective aperture of the antenna in square meters (m^2).

In theory at least, the brightness temperature of a radiometer transducer (the antenna in the case of a radio telescope) is exactly equal to the absolute temperature of the source. Ideally, a star burning at 8600°K produces an equivalent brightness temperature of 8600°K at the antenna. Assuming an antenna has an effective aperture of 10 m^2, Equation 5-1 shows that an 8600°K source should produce a brightness flux density of 2.4×10^{-24}°K/m^2 at the antenna terminals.

What does this mean to an amateur radio astronomer? Probably nothing.

To make matters worse, the usual means of specifying line losses, amplifier gain, and noise levels have to be translated into brightness units, too. The pressure has become more than electrical engineers are willing to bear in the past few years, and the whole notion of brightness temperature is, for the most part, giving way to more familiar terms. Equation 5-1 has some value, but amateurs tend to shy away from it.

The fundamental equation of radio astronomy can be stated in more familiar terms;

$$S = \frac{P_a}{A_e B} \tag{5-2}$$

where

P_a = the power at the antenna terminals, in watts
A_e = the effective aperture of the antenna, m^2
B = the predetection bandwidth of the system, in hertz (Hz)
S = the signal flux density in watts per square meter per hertz ($W/m^2/Hz$).

Example 1: Suppose a receiver having a bandwidth of 10 kHz is capable of detecting signals as low as 10^{-15}W. If the antenna aperture is 5 m^2, what is the minimum flux density the system can detect? From Eq. 5-2:

$$S = \frac{10^{-15}}{(5)(10\times10^3)} = 2\times10^{-20}\ W/m^2/Hz$$

Equation 5-2 is worthy of special study because it embodies the main principles of radio telescope design. The equation is especially enlightening if it is rearranged to show $P_a = SA_eB$. This version shows that the signal power available at the antenna terminals is directly proportional to the flux density of the source, the aperture of the antenna, and the rf bandwidth of the system. It is perhaps obvious that stronger radio sources should produce more power at the antenna terminals than weaker sources can. But it is important to note that increasing the effective aperture or working area of the antenna produces greater power levels for a given source intensity. In light of this fact, it is little wonder radio astronomers are always striving for larger antennas.

By the same token, increasing the antenna and receiver bandwidth makes it possible to detect weaker sources at a given level of power sensitivity. And to sum up all of these facts: The greater the power-level sensitivity of a system, the more sources it can detect.

The remainder of this chapter uses Eq. 5-2 as a starting point for developing the main design equations for modern radio telescope systems. It is possible to assemble a working system without the help of the equations—just as it is possible to put together an audio amplifier without resorting to circuit calculations. The design equations that appear in the following sections, however, enable the experimenter to make wiser choices of equipment and predict with greater reliability the chances of success.

BASIC ANTENNA DESIGN PARAMETERS

Equation 5-3 relates the flux-unit intensity of a radio source to the amount of power, expressed in watts, developed at the antenna terminals. This particular equation differs from the one in the previous section only by a constant multiplier of 10^{26} which converts flux density S into a term radio astronomers call *standard flux units, F.U.*

$$\text{F.U.} = \frac{P_a \times 10^{26}}{A_e B} \qquad (5\text{-}3)$$

Determining the flux-unit strength of a signal is thus a matter of determining the power available at the antenna terminals, given a certain antenna aperture and predetection bandwidth. Under normal operating conditions, the experimenter records the antenna power in one way or another. Fortunately, most receiver systems have a square-law detector that transforms rf and signal power into a proportional voltage. The signal voltage measured after the receiver's detector stage is thus a very good indication of the amount of power at the antenna.

Two of the components in Eq. 5-3—the antenna aperture and system's rf bandwidth—are normally constant once the system is set up and running. That leaves only the power at the antenna terminals P_a to be determined as a routine part of a radio astronomy project.

The receiver's bandwidth can be determined either from the manufacturer's specifications or by direct measurement, and the antenna bandwidth is a characteristic of the antenna type. The term B in these equations must be equal to the lesser of the two bandwidths. If, for instance, the antenna and receiver have bandwidths of 10^5 Hz and 4 MHz, respectively, B should be equal to 10^5 Hz.

It is far more difficult to arrive at an accurate figure for the antenna's effective aperture A_e. The aperture is indeed proportional to the physical area of the antenna, but the ratio of effective-to-physical area is a rather elusive one. The effective area of an antenna is usually less than the physical area—an amateur experimenter can count on it being so. In spite of the vague ralationship between the *effective* and *physical* area of an antenna, it is possible to make one very broad generalization: *effective* and *physical* area tend to approach a 1:1 ratio as the resonant wavelength shortens. In the upper UHF region, for instance, a well designed antenna might have an effective area that is about 80–90% of its physical area. In the gigahertz range, the effective area can actually exceed the physical area by as much as 10%.

Equation 5-4 can serve as a guide to estimating the effective area or aperture of a given antenna.

$$A_e = \frac{D\lambda^2}{\pi 4} \tag{5-4}$$

where

A_e = the effective aperture of an antenna, m^2
D = the directivity of the antenna (a unitless term)
λ = the antenna's design-center wavelength, m.

The difficulty with Eq. 5-4 is that it, too, requires a knowledge of a term that is hard to evaluate—directivity D. The directivity of an antenna is defined as its ideal power gain expressed as the ratio of *power out* to *power in*. Although it is often difficult to determine the true directivity of an antenna, it is generally better known than the aperture; and Eq. 5-4 can be used to estimate the effective aperture fairly well.

The gain of an antenna or general antenna classification is normally expressed in decibels of power gain. Since the directivity term in Eq. 5-4 requires a straightforward ratio rather than decibels of gain, it is necessary to convert the decibel figure.

Equation 5-5 makes the necessary conversion:

$$D = \frac{\text{antilog } G_p}{10} \tag{5-5}$$

where

D = the antenna's directivity

G_p = the power gain of the antenna, decibels (dB)

Example 2: Suppose an experimenter wants to determine the effective aperture of an antenna having a typical gain of 12 dB at a frequency of 100 MHz.

The ultimate solution of this problem rests with Eq. 5-4. The antenna directivity *D* and the wavelength λ must be determined first, however. By Eq. 5-5, the directivity is:

$$D = \frac{\text{antilog } 12}{10} = \text{antilog } 1.2$$

$$\approx 16$$

The wavelength can be found by the equation:

$$\lambda = \frac{300}{f}$$

where f is frequency, megahertz. Thus,

$$\lambda = 300/100 = 3 \text{ meters.}$$

Using this information in Eq. 5-4:

$$A_e = \frac{(16)(3)^2}{\pi\, 4}$$

$$\approx 11.5 \text{ square meters}$$

For the sake of computational simplicity, Eqs. 5-4 and 5-5 can be combined into one. After evaluating the constants, the equation becomes that of Eq. 5-6.

$$A_e = \frac{7.17 \times 10^3 \text{ antilog } (G_p/10)}{f^2} \tag{5-6}$$

where

A_e = the effective aperture of the antenna, m^2

G_p = the antenna gain, dB

f = the antenna's design resonant frequency, MHz

Example 3: What is the effective aperture of an antenna cut for 800 MHz if it has a characteristic gain of 8 dB? From Eq. 5-6:

$$A_e = \frac{7.71 \times 10^3 \text{ antilog } 0.8}{800^2}$$

$$= \frac{7.17 \times 10^3 \ (6.3)}{800^2}$$

$$= 0.07 \text{ m}^2$$

Summarizing the information presented thus far in this section: An experimenter uses Eq. 5-3 to routinely determine the flux-unit signal strength of a radio source. Using that particular equation requires a knowledge of the antenna's effective aperture and the system's predetection bandwidth. The bandwidth of a system is generally known with a suitable degree of accuracy. The aperture can be closely approximated by Eqs. 5-4 and 5-5, or by Eq. 5-6 alone.

Besides being a valuable equation for amateur radio astronomy, Eq. 5-3 can be rearranged to serve as a system design equation. Solving Eq. 5-3 for antenna power P_a, it becomes:

$$P_a = (F.U.) A_e B \times 10^{-26}$$

And by substituting Eq. 5-6 for the A_e term, the result is Eq. 5-7.

$$P_a = \left(\frac{7.17 \text{antilog } \frac{G_p}{10} B(F.U) \times 10^{-23}}{f^2} \right) \qquad (5\text{-}7)$$

where

P_a = the signal power available at the antenna terminals, *W*

G_p = the antenna gain, *dB*

B = the system's predetection bandwidth, *Hz*

F.U. = the standard flux-unit signal strength at the antenna

f = the design resonant frequency of the antenna, *MHz*.

Example 4:

An experimenter wants to be able to detect radio sources having flux-unit strengths of 1000 or more. If he uses a commercial FM receiver having a bandwidth of 100 kHz and an antenna having a typical gain of 15 dB at 110 MHz, what must his receiver be able to detect in terms of signal power? From Equation 5-7:

$$P_a = \left(\frac{7.17 \text{antilog} \frac{15}{10} (100\times10^3)(1000)\times10^{-23}}{110^2} \right)$$

Since antilog 1.5 = 31.6

$$P_a = 18.7\times10^{-18}\text{W}$$

This experimenter's receiver system must be capable of amplifying 18.7×10^{-18}W to a level that is compatible with his recording equipment.

A standard *F.U.* value of 1000 is about the highest value one can use to make a project worthwhile. As shown in Appendix B, only five radio sources have intensities of 1000 *F.U.* or greater.

Example 5:

Suppose the experimenter in the previous example decides he would like to lower his minimum detectable *F.U.*. figure to 800 *F.U.* Using the same antenna and receiver, he most likely has to add an rf preamplifier. If he uses a preamplifier with a rated power gain of 5 dB, what will be the signal power level at 800 *F.U.*? (Assume all other parameters are the same as those in Example 4.)

The preamplifier increases the effective power gain of the antenna G_p by 5 dB. In other words, $G_p = 15+5 = 20$ dB. Antilog 20/10 = 100. Using that figure for gain and the lower *F.U.* base of 800 in Eq. 5-7:

$$P_a = 47.4\times10^{-18}\text{W}$$

Using the rf preamplifier not only lowered his minimum detectable source to those having 800 *F.U.* or better, it delivered more power to the receiver.

RECEIVER DESIGN PARAMETERS

Once an experimenter gets a good idea of how much power his receiver must be able to amplify, he should turn his attention to the task of determining whether or not the receiver can do the job. The key specification used in this book, as far as the subject of receivers is concerned, is sensitivity.

Receiver sensitivity is normally specified in terms of the number of microvolts at the input that produces an output signal 10 dB above the internally generated noise. The expression "10 dB above" is figured according to the ratio: $(S+N)/N$, where S and N represent signal and noise power, respectively.

Fortunately for amateur radio astronomers, their *signal-plus-noise-to-noise* ratios can be much worse than normal communications technology allows. In fact, an extraterrestrial source creating a signal that is 10 dB above the receiver noise level would be a remarkable source indeed. Only the most powerful sources such as the *Sun* and Jupiter produce output signals that equal the system noise, and most amateur radio astronomers learn to be quite comfortable with signal levels that are only a fraction of the normal noise level from their system.

The point of all this is that a receiver having a specified sensitivity of 1 μV is, in actual practice, capable of detecting signals on the order of 0.01 μV. Of course, the recording device will not exactly drive itself into saturation with a 0.01 μV signal at the antenna, but the change in noise level will be discernible.

This section deals with a technique for determining the lowest possible signal power the receiver can reliably reproduce, given the receiver's rated sensitivity. If, for instance, an experimenter knows his receiver has a sensitivity of 1 μV, he would like to find out whether or not he can detect signals from 1000 *F.U.* sources.

The first step in the procedure is to determine the smallest possible change in voltage the readout device can reliably reproduce. The smallest detectable change is closely related to the instrument's precision; and if the manufacturer's specifications show a precision of ± 0.1V, for example, the experimenter can rightly assume the smallest detectable signal is on the order of 100 mV (0.1V).

The second step is to determine the normal noise voltage at the readout device. Replace the antenna with a fixed resistance having a value equal to that of the antenna impedance and measure the output noise level. For most applications, the noise level should be read with the receiver gain at its maximum level.

Applying Eq. 5-8 leads to a figure representing the smallest possible antenna power level the system can reliably reproduce.

$$P_{ar} = \frac{(0.11 \Delta S_m V_{ar})^2 10^{-12}}{N_0^2 Z_a} \qquad (5\text{-}8)$$

where

P_{ar} = the minimum detectable antenna power, W

ΔS_m = the smallest detectable change in output potential, V

V_{ar} = the receiver's sensitivity, μV

N_0 = normal output noise level, V

Z_a = the antenna impedance.

Example 6:

Suppose an experimenter wants to use a receiver having a rated sensitivity of 1 μV. The output noise level, with an equivalent resistance replacing the antenna, is 0.7V; and the smallest detectable change in voltage on that scale is 20 mV. If the antenna has an impedance of 52 ohms, what is the smallest antenna power level he can detect? By Eq. 5-8:

$$P_{ar} = \frac{[(0.11)(0.02)(1)]^2 \times 10^{-12}}{(0.7)^2(52)}$$

$$\simeq 0.19 \times 10^{-18} \text{W}$$

This means the smallest radio source he can detect is one that produces about 0.2×10^{-18}W at the antenna terminals. The experimenter in Example 4 found that his antenna was able to transform a 1000 *F.U.* source into about 19×10^{-18}W. The receiver specified in Example 8 is adequate for the job. The experimenter will, in fact, be able to detect even weaker signals and compensate for unaccountable line losses.

COMBINING ANTENNA AND RECEIVER PARAMETERS

The antenna is generally the unique portion of a radio telescope installation—it is the one part that usually has to be built up from scratch. On the other hand, the experimenter is generally bound by practical considerations to use a receiver he has at hand. Under such circumstances, the experimenter should apply Eq. 5-8 to determine how well his receiver can perform, then design or select an antenna that will produce the required amount of signal power for the kinds of projects he wants to perform.

A second situation arises where the experimenter already has an antenna set up and ready to go, and he has an opportunity to select a receiver. In this case, he determines the smallest antenna power output for the sources he wants to observe, then selects a receiver having the required bandwidth and sensitivity.

The third and ideal situation is one where the experimenter can select both the antenna and receiver on the basis of the weakest sources he wants to observe.

All three conditions are considered in the remaining parts of this chapter. A careful study of the examples can lead to a better understanding of radio telescope engineering as well as practical information concerning the selection of equipment.

Selecting an Antenna for a Given Receiver

Example 7: An experimenter has access to a communications receiver having a bandwidth of 10 kHz, a sensitivity of 1 μV, and a tuning range in the 18 to 30 MHz bands. Tapping the audio output from the volume control, he finds a normal noise level of 1V. If he wants to use the system to observe Jupiter's 10^5 *F.U.* signals at 24 MHz, what antenna signal power is required to transform a 10^5 *F.U.* signal into a 2V change on his readout device? Assume the antenna will have an impedance on the order of 52 ohms.

Equation 5-8 is appropriate in this case, providing the minimum detectable output voltage ΔS_m is replaced with the minimum output voltage the experimenter wants at 10^5 *F.U.*

$$P_{ar} = \frac{[(.11)(2)(1)]^2 \times 10^{-12}}{(1)^2(52)}$$

$$= 9.3 \times 10^{-16} W$$

To see a 2V increase in the output voltage, then, the antenna signal must increase by 9.3×10^{-16}W. The next problem is to determine the antenna gain required to produce that power level when it sees 10^5 *F.U.*

Rearranging Eq. 5-7 to solve for antenna gain:

$$G_p = 10 \log \frac{P_a f^2 \times 10^{23}}{7.17B(F.U.)}$$

$$= 10 \log \frac{9.3 \times 10^{-16}(24)^2 \times 10^{23}}{7.17(10^4)(10^5)}$$

$$= 10 \log 7.47$$

$$= 8.7 \text{ dB}$$

To do the job, then, the antenna should have a power gain that is close to 10 dB. Perhaps a 5- or 6-element yagi antenna would be the best all-around choice.

Selecting a Receiver for a Given Antenna

Example 8:

An amateur radio astronomer has access to a commercial multiband antenna having a listed UHF gain of 15 dB. The antenna's terminal impedance is about 72 ohms. Determine the required receiver sensitivity, given the following conditions: The receiver bandwidth can be as high as 5 MHz at an operating frequency of 800 MHz, and he wants a 1000 *F.U.* source to display a 3:100 increase over normal noise voltage.

According to Eq. 5-7, the antenna will deliver the following power level at 1000 *F.U.*:

$$P_a = \left(\frac{(7.17 \text{ antilog} \frac{15}{10})(5 \times 10^6)(10^3) \times 10^{-23}}{800^2} \right)$$

Since antilog 1.5 = 31.6

$$P_a = 1.77\times10^{-17}\text{W}$$

The next part of the operation is to solve Eq. 5-8 for the receiver sensitivity at that power level. Rearranging Eq. 5-8:

$$V_{ar} = \frac{N_0\times10^6\sqrt{P_aZ_a}}{0.11\ S_m}$$

$$= \frac{100\times10^6\sqrt{1.77\times10^{-17}(72)}}{0.11(3)}$$

$$\simeq 11\ \mu\text{V}$$

INCREASING SYSTEM SENSITIVITY

Equation 5-8 shows that the smallest detectable power level at the receiver's input terminals is proportional to the square of the smallest readable output voltage on the readout device S_m divided by the average output noise N_o level, or S_m/N_o^2. This section describes some useful techniques for increasing the S_m/N_o ratio and thus greatly enhancing the minimum detectable antenna signal.

As far as Eq. 5-8 is concerned, S_m/N_o is fixed by the electrical specifications of the receiver and the precision of the readout device. The discussions and examples related to Eq. 5-8 tell the user to either measure the N_o and determine the smallest voltage increase he can read on the output device, or fix an S_m/N_o ratio that leads to a minimum detectable signal that is no less than about 10% of the average noise level ($S_m/N_o = 0.1$). Adjusting the ratio from 1:1 to 1:10 increases the system's *effective power sensitivity* by a factor of 100.

The system's inherent noise level N_o is indeed fixed by the electrical characteristics of the receiver system and readout device. It is possible, however, to make the S_m figure much smaller by further modifying the signal at the output device and by manipulating the recorded data:

$$S_m' = \frac{S_m}{\sqrt{tn}} \qquad (5\text{-}9)$$

where

S_m' = the adjusted minimum detectable change in output signal, V

S_m = the smallest detectable change in output as read directly from the readout device, V

n = the number of records averaged together.

For the purpose of this discussion, the system's postdetection integration time is roughly equal to the time constant of the readout device. If, for example, the time constant is equal to 2 sec, the term t in Eq. 5-9 can be set to 2.

The time constant of the system tends to increase the ultimate sensitivity by virtue of the fact that noise levels tend to remain fairly constant and random, while signal levels tend to increase rather slowly over a relatively long period of time. The system's time constant, in other words, filters out unchanging noise levels, but lets the readout device respond to overall increases that occur when a radio source passes through the antenna beam.

The number n of records averaged also contributes to the system's overall sensitivity. The basic idea is to record raw signal data from the same source several times, then average the readings. System noise is random in nature, while signal-level increases from radio sources are not. Averaging several recordings from the same source thus tends to eliminate the random components, while leaving the signal responses intact.

Optimizing Integration Time

Figure 5-2 shows an output integration circuit that lets the user vary the postdetection integration time between 0.5 and 5 sec. The circuit is a simple low-pass filter that has a time constant roughly equal to the product of R2 and C_f, where C_f is the value of the integrating capacitor (C2, C3, C4, or C5) in farads. The input side of the circuit can be attached to either the receiver's volume control terminals or the leads on the loudspeaker.

Probably, the best way to determine the optimum time constant for a given system and project is by trial and error. Since the Sun is our most powerful and reliable natural radio

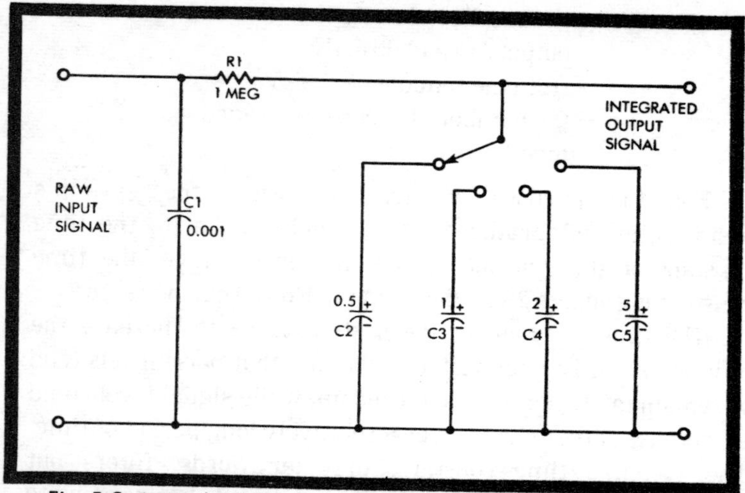

Fig. 5-2. A simple circuit for integrating signals from the receiver.

source, set up the system to detect the Sun and adjust the time constant of the integrator circuit for the best all-around response.

Example 9: If an experimenter decides he can reliably see a 1V increase in signal level "riding" on 4V of noise with his readout device, how much increase in sensitivity can he expect if he sets the postintegration time of the system to 5 sec?

The given information sets the basic S_m/N_o figure from Eq. 5-8 to ¼, or 0.25. Equation 5-9 shows that the adjusted signal S_m' is equal to $1/\sqrt{5}$, or 0.45—assuming $n = 1$ for on-line readings. The user can thus expect to observe a 0.45V signal on top of a normal 4V noise level.

Before placing the 5-second integration time into the output circuit, the $(S_m/N_o)^2$ term for Eq. 5-8 was equal to $(0.25)^2$ or about 0.06. The adjusted ratio is 0.45/4, and squaring that term leads to 0.11^2 or 0.01V. Adding the 5-second integrator thus increases the power sensitivity of the receiver system 5 times.

It is no coincidence that the increase in power sensitivity in Example 9 is equal to the integration time of the system. The increase in power sensitivity, as expressed in Eq. 5-8, is indeed equal to the product of t and n.

It would be nice if it were possible to increase the integration time indefinitely, or at least to a point where signal changes on the order of 10 minutes or more were all that could register on the readout device. Practical considerations—namely, the availability and reliability of high-value capacitors—limit the upper value of integration time.

Limitations of the Signal-Averaging Process

The power sensitivity of the radio telescope system effectively increases in proportion to the number of records averaged. It is easy to misinterpret this fact, however; and an experimenter must understand the practical limitations and significance of the signal-averaging procedure.

Since the signal-averaging process takes place after gathering data several times from the same source, there is no way to say the procedure has any direct bearing upon the actual system sensitivity. Signal-averaging takes place "after the fact," and there is no point in trying to average the noise out of recordings where the signal is too small to measure in the first place.

Speaking in very general terms, the signal-averaging process cleans up signal responses that already exist on the recordings. Rather than actually increasing the output signal level, the process lowers the system's noise level; and the practical limitation of the process is the amount of time and patience the experimenter is willing to put into calculating the averages.

INCREASING SYSTEM RESOLUTION

The technical discussions to this point deal mainly with the signal strength of a radio source and the overall sensitivity of a radio telescope system. This section takes up the matter of system resolution—the system's ability to distinguish a source covering 30 minutes of arc from one covering perhaps 1 minute.

The simple radio telescope system described in previous sections of this chapter is not capable of providing very much information about the angular size of a radio object. For the

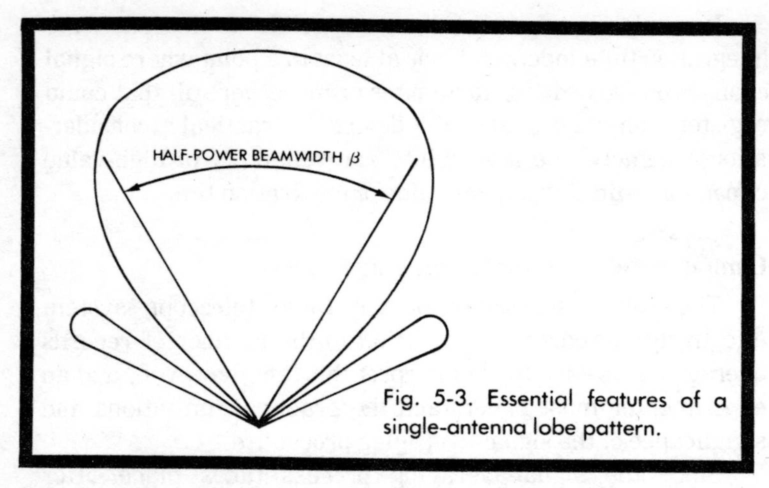

Fig. 5-3. Essential features of a single-antenna lobe pattern.

most part, a radio galaxy and our own Sun appear to occupy the same angular space. It is possible, however, to construct a radio telescope system that gives the observer a much better notion of a radio object's apparent size.

Before launching into a discussion of a high-resolution system, it is important to look at the normal kind of system from a different point of view. Improving the resolution of a radio telescope is strictly a matter of reworking the antenna system—the electrical characteristics of the receiver and readout system have little to do with the overall resolution.

Radiometer Antenna Patterns

A radio telescope system composed of a single antenna, a receiver, and a readout device is technically called a *radiometer*. All of the discussions in this chapter have, to this point, concerned radiometers.

Figure 5-3 shows a typical beam pattern for a radiometer telescope. The beamwidth between half-power points is determined mainly by the antenna design. The actual beam pattern for any kind of antenna is determined by a complex combination of basic design, distance from the Earth, its proximity to any kind of large conductive materials, and so on. For the purposes of discussion, it is helpful to consider a semi-ideal beam pattern showing only one major lobe and a pair of side lobes.

Figure 5-4 illustrates an idealized response curve for a small radio source passing through the antenna beam. Assuming the source has a fairly constant energy output, the reading at the telescope output shows a small rise and fall in power level, one large change in power level as the object passes through the major lobe, and a smaller change that occurs as the object moves through the second minor lobe.

The width of the tracings is determined by the antenna's characteristic beamwidth. If, for example, the antenna has a major lobe with a 30° spacing between half-power points, a source will pass through the beam in about 2 hours. (Recall that celestial bodies move across the sky at an apparent rate of about 15° per hour.) A very small radio source such as *Cygnus A* crosses the beam in about the same amount of time as a larger source like the Sun. The angular diameters of these two sources are vastly different, but the resolution of a typical amateur radiometer is much too poor to show any difference. Only the galactic plane or Milky Way is large enough to show up as a source that is any larger than a star on a simple system.

The only way to improve the resolution of a radiometer telescope is by increasing the directivity of the antenna system; and that means increasing its physical size many

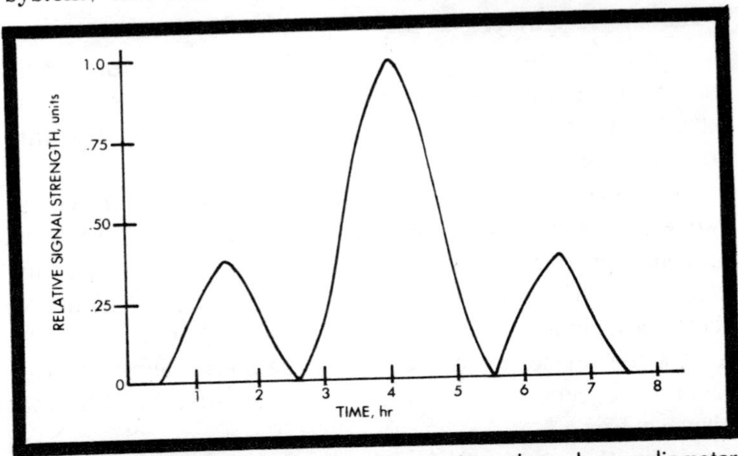

Fig. 5-4. Signal response of a source passing through a radiometer beam—minor responses are caused by side lobes.

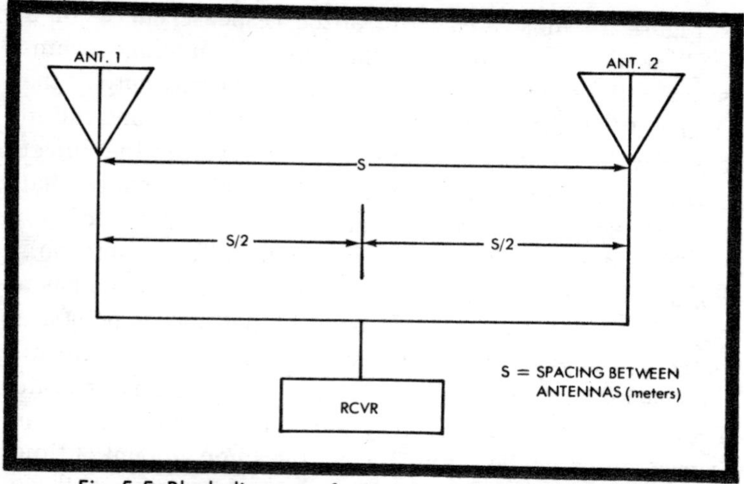

Fig. 5-5. Block diagram of a typical interferometer system.

times over. Professional radio astronomers who can gather sufficient funds build high-resolution radiometers having antennas with immense apertures. This is a brute-force technique that is far beyond the means of most amateurs. A more satisfactory way to increase the overall resolution of a small radio telescope system is by converting it from a radiometer to an *interferometer*.

Radio Interferometers

The only difference between the block diagram for a basic radiometer and an interferometer radio telescope is that the latter uses a pair of identical antennas separated by a certain horizontal distance.

Whenever a radio signal from a single source arrives at two widely spaced antennas, the combined signal from the antennas appears as an interference pattern—one showing periods of aiding and canceling. If the antennas are identical, their individual signals should be identical in all respects except the phase angle. And when the source emits a wide band of frequencies, the resulting lobe pattern resembles that shown in Fig. 5-6.

Using the interferometer technique effectively breaks up a large central lobe into a number of smaller ones. As illustrated

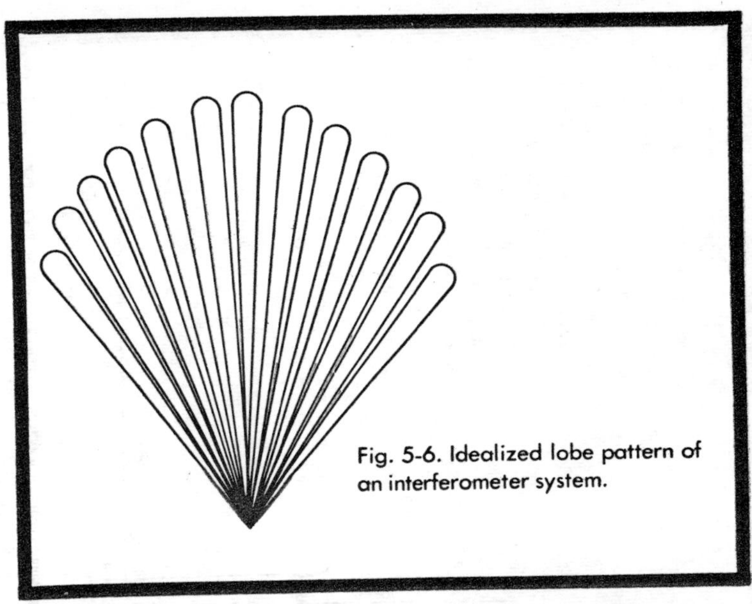

Fig. 5-6. Idealized lobe pattern of an interferometer system.

in Fig. 5-7A, a very small source crossing through the antenna pattern generates a series of smaller response curves rather than a single large one. Figures 5-7B and 5-7C show how the system responds to radio sources having apparent angular diameters that are large compared to the beamwidth of the interference globes.

Equation 5-10 shows the close relationship between the width of the interference lobes and the physical spacing of the two antennas.

$$\beta = \frac{1.7 \times 10^4}{fs} \qquad (5\text{-}10)$$

where

β = the beamwidth (in degrees) between half-power points of the interference lobes

f = the center frequency of the antennas, MHz

s = the spacing of the two antennas, m

Suppose, for instance, a system is designed for operation at 100 MHz and the interferometer antennas are separated by

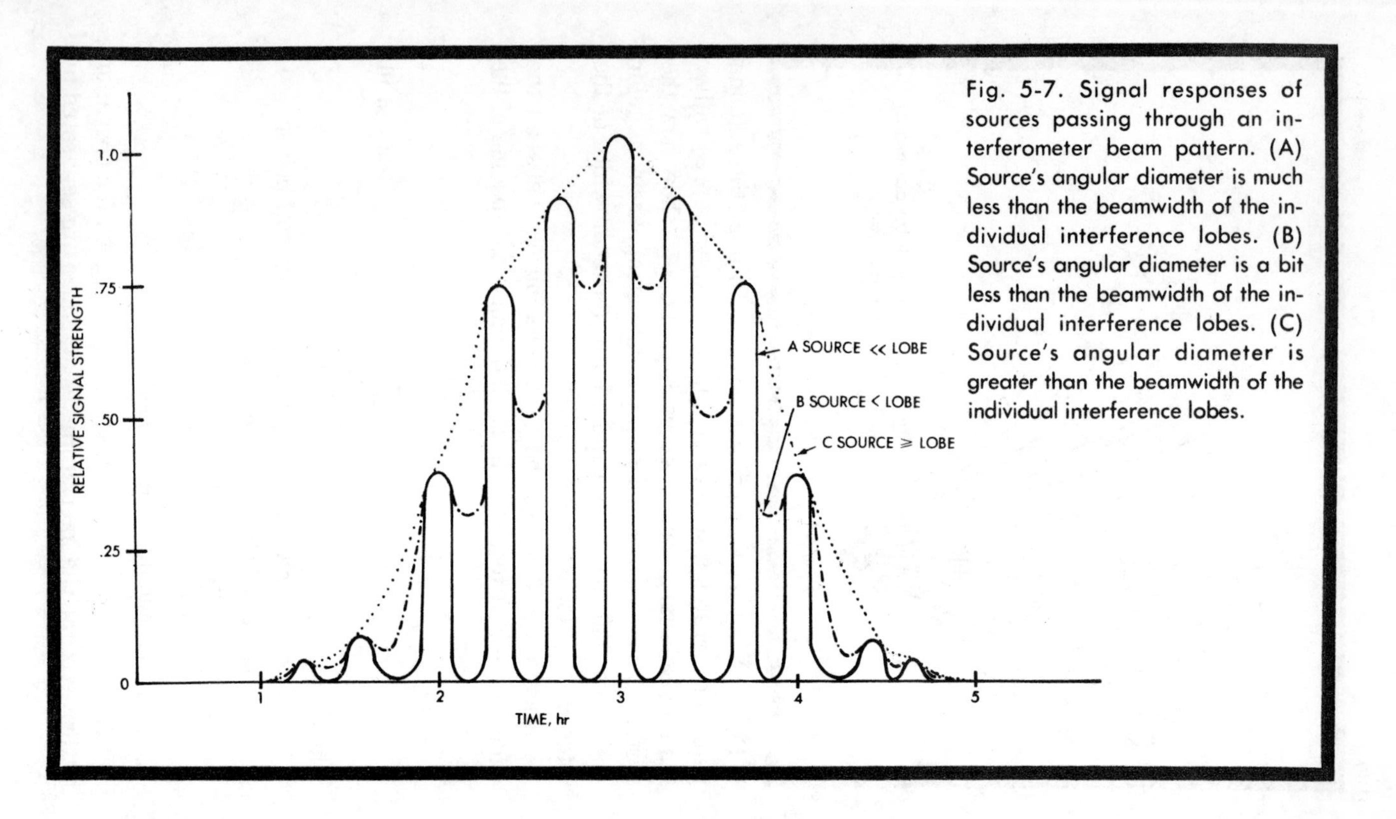

Fig. 5-7. Signal responses of sources passing through an interferometer beam pattern. (A) Source's angular diameter is much less than the beamwidth of the individual interference lobes. (B) Source's angular diameter is a bit less than the beamwidth of the individual interference lobes. (C) Source's angular diameter is greater than the beamwidth of the individual interference lobes.

a distance of 40m. The beamwidth of the interference lobes is thus:

$$\beta = \frac{1.7\times10^4}{100\times40} = 4.25°$$

The simple interferometer system in this case is capable of resolving radio objects down to about 4°15′. Considering one of the antennas used alone might have a beamwidth of 30° or more, this represents a substantial gain in resolution. Of course, the system sensitivity also increases because the effective antenna aperture is doubled by using two identical antennas.

One of the most important ideas suggested by Eq. 5-10 is that the resolution of an interferometer radio telescope increases with the spacing between the antennas. There are, however, two practical considerations that place serious limitations upon the antenna spacing: the amount of physical space available to accommodate the antennas, and the losses in the transmission lines interconnecting them.

The transmission lines running from the antennas to the receiver must be equal in length *and* impedance. The best way to insure these conditions is to buy a length of transmission line that is equal to the antenna spacing, cut it at its center, and make a tee connection from that point to the receiver.

It is tempting to use rf preamplifiers at the antennas to boost the signal level and thereby make up for losses in very long transmission lines. Since no two amplifiers are exactly alike, however, it is impossible to achieve a good interference pattern from their combined outputs. Antenna conditions must be as close to identical as possible, and preamplifiers at the antenna can scramble the phase and amplitude relationships beyond hope. A single preamplifier can, however, be properly installed at the center connection point where the two transmission lines meet.

Amateur Radio Astronomy Projects

Radio astronomy is a new science as far as amateurs are concerned; as such, it has very few established rules and even fewer recipe-type projects a beginner can conduct with absolute assurance of success. And what's more, there is no equipment yet available that is specifically designed for amateur radio astronomy work.

The present-day situation of amateur radio astronomy has two sides, however. On the negative side, the lack of information, established procedures, and specialized equipment makes it very difficult for an unskilled experimenter to enter the game. On the positive side, amateur radio astronomy offers unparalleled opportunities for devising new projects and circuits that can be of immense value to others.

Amateur radio astronomy is a demanding activity for anyone because it calls for a peculiar combination of patience, theoretical knowledge, practical know-how, and imagination. For those who have or can develop these traits along the way, amateur radio astronomy can be an especially rewarding pursuit.

This chapter deals with amateur radio astronomy projects and procedures in very general terms. The main point is to show a beginner what he can do once he selects and assembles an installation. Most of the projects described here are modified versions of the work professional radio astronomers are doing. A few of the suggested projects—and they are suggestions rather than explicit instructions—are those the professionals haven't had time to try for themselves. Professionals have chosen projects that promise the biggest payoffs in terms of results—namely, those projects involving searching and mapping the sky in the microwave range. There is an almost limitless variety of different projects and frequency ranges the pros haven't had time to explore. And therein lies the real opportunity for amateurs.

BASIC PROCEDURES AND EQUIPMENT

Although amateur radio astronomy is wide open with respect to the selection of specific projects and equipment, there are some inescapable general principles that make radio astronomy something quite different from optical astronomy and radio communications technology. One set of principles concerns the hardware elements of a radio telescope system, and another takes into account basic celestial mechanics and the nature of radio emissions from celestial bodies. A third set of principles—not really unique to radio astronomy, but vital to its success—concerns scientific methodology: the procedures and thinking processes that set apart serious investigators from science quacks.

General Hardware

Figure 6-1 shows a block diagram of a basic radio telescope system. The diagram is universal in the sense that no radio telescope—no matter how small and simple, or large

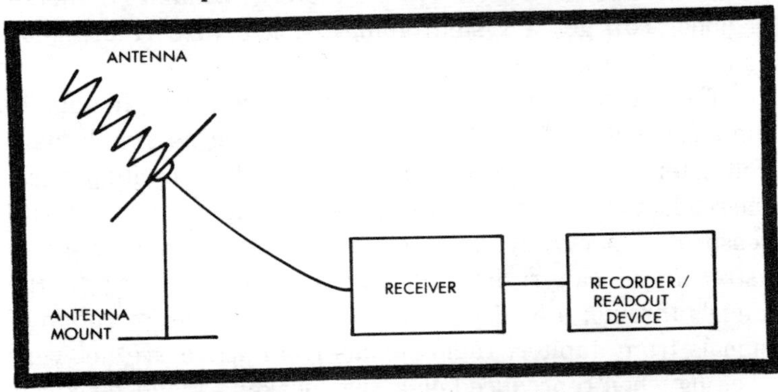

Fig. 6-1. Block diagram of a typical radio telescope system.

and complex—can be complete without having all of these components in one form or another.

Any radio telescope system includes an antenna and its mount, a receiver, feedline running between the antenna and receiver, and some sort of readout or signal recording device. Perhaps anyone knowledgeable in the ways of communications electronics finds the purpose of each of these elements in the system to be rather obvious: the antenna picks up radio signals from remote sources; the antenna mount steers the antenna; the feedline carries the antenna signal to the receiver; the receiver tunes, amplifies, and detects the signal; and the recording system displays or records the information in some meaningful form. The special requirements imposed upon these essential components are the factors that usually require some extra discussion.

The antenna, for example, must have a higher power gain than is normally required for communications work, and it must be highly directive. Only antennas with gain figures in excess of 10 dB are worth considering for radio astronomy work. The antenna should also have a wide bandwidth—something on the order of 2 MHz.

As far as the receiver is concerned, it must have high gain, low noise, good stability, and wide bandwidth. These particular requirements are almost self-contradictory, and run-of-the-mill radio construction projects seldom meet them. Fortunately, VHF and UHF technology has developed to a point where suitable receivers are readily available. In fact, a beginner can get a system going using a moderately good commercial FM receiver.

The data readout or recording device is the final stage in the electrical system. Such a device can be as simple as a voltmeter or as complex as a chart recorder. In either case, the readout or recording device must provide information with reasonable accuracy and reliability. Few amateur radio astronomers can resist the temptation of listening to the signals through a loudspeaker. With the notable exception of signals from Jupiter, radio signals from space are not very exciting when reproduced on a loudspeaker; a loudspeaker is nevertheless a popular kind of "readout" device.

Most amateur radio astronomers find they spend a surprising amount of time and effort designing, constructing, and working the mechanical bugs out of the antenna mounting. Electronically oriented experimenters sometimes tend to think of the antenna mount as a minor mechanical nuisance; but the ultimate success or failure of a radio telescope installation rests heavily upon the quality of the antenna mounting.

Basic Radio Astronomy Projects

Projects suitable for amateur radio astronomy fall into the general categories of studying discrete sources and mapping. As the name implies, mapping projects concern the process of making maps of the radio sky. Since professional radio astronomers have been carrying out mapping projects for a number of years, one might think a mapping project would be little more than a technical exercise. This is not necessarily the case, however.

Professional radio astronomers have indeed made a number of reasonably complete radio maps of the sky, but only at a few different wavelengths. Certain thermal radio sources tend to emit different amounts of energy at different frequencies, and the sky as observed at 1200 MHz is bound to be different from that "seen" at 500 MHz. For the next few years at least, every amateur radio astronomer has an opportunity to map the sky as it has never been seen before—at frequencies professional and other amateurs have not had the time nor the inclination to try.

Closely related to straightforward mapping projects are systematic searches for undiscovered radio sources and anomalous emissions. The procedures in such cases are similar to those used in mapping projects, but the intent is altogether different. Once an experimenter is familiar with the radio sky at a certain frequency, he can plan a search for sources that haven't been cataloged before. The discovery of *pulsars* was accidental, but it represents the sort of discovery open to anyone searching for the unusual in space.

The second major class of amateur radio astronomy projects is that of studying discrete sources. Discrete-source studies are different from mapping projects in that the former concentrates upon a single object or point in the sky. There is a wide variety of discrete-source studies of interest to radio astronomers, and some of them can be quite original. A brief list of such studies would include observations of lunar occultations, sunspots, eclipses, variable stars, and occultations of Jovian moons.

A beginner would do well to play around with his system before launching any formal radio astronomy projects. The informal projects described in the following section can test the operation of a new installation, give the user a feeling for operating the equipment, and provide some positive and encouraging results.

INFORMAL PROJECTS

Many amateur radio astronomers prefer to turn on their equipment now and then and simply play around. These people find the greatest satisfaction in spotting a few well known radio sources, recording some data now and then, and trying to identify the nature of signals they receive at times most convenient to them.

It is quite natural to want to get some results after spending weeks or months planning and assembling a new system. A good first test of a new system is to aim the antenna at the Sun and compare the signal with that found by pointing the antenna away from the Sun. If the Sun is not in the sky at the time, the Milky Way or one of the other powerful sources such as Cygnus A can serve just about as well. A system that doesn't respond to one of these sources won't respond to any other signals from space.

Once the experimenter knows he can see the most powerful radio sources, he can fix the antenna's azimuth on the meridian, set the altitude to some active part of the sky such as the ecliptic, and record the signal output at intervals of 30 minutes or so. As the Earth rotates, it carries radio sources through the antenna beam, causing changes in the output level. By noting the antenna's azimuth and altitude headings, and the exact time the output shows a high signal level, the experimenter can compute the right ascension and declination of the source. After that, he can identify the source by comparing his calculated coordinates with a standard celestial map or radio map of the sky. An experimenter can also predict the time of transit of a known source and watch for its signal on his readout device.

It is always thrilling to watch the system respond to a predicted change in signal level; but it is even more exciting to find an unexpected change and later associate it with a source previously overlooked.

An amateur working on a casual project level can treat radio astronomy as the "shortwave listener" treats ham radio—keeping a list of confirmed radio sources and looking around for new ones. Once amateur radio astronomy becomes an established hobby, it is quite likely amateurs will run source-spotting competitions and award certificates of achievement.

Equipment for Informal Projects

Informal radio astronomy projects can be the least demanding kinds of projects in terms of the amount of equipment required. A commercial FM receiver works quite well for VHF work and a communications receiver (usually with the help of a preamplifier) can do the receiving job in the decameter range.

The antenna requirements for informal projects are much the same as for the more serious undertakings, but the mounting schemes can be much simpler. A long beam antenna with an altitude mounting is good for informal studies in the VHF and decameter ranges.

The recording system for informal radio astronomy work can be an ordinary VTVM connected across the receiver's volume control or at the loudspeaker terminals Using a moderately good receiver and an antenna system with a gain of 10 dB or so, the experimenter might expect to see the normal noise voltage from the receiver increase by about 10% as the Sun or one of the other major radio sources crosses through the beam.

Precautionary Notes

Informal radio astronomy projects are normally carried out using the simplest kinds of equipment and the most casual kinds of engineering and scientific procedures. With simplicity and casualness being the hallmarks of informal projects, the beginner should be forewarned that his results might be quite exciting, but not very reliable with respect to new discoveries.

In connection with the more formal projects later described in this chapter, it is absolutely essential to carefully verify the presence of every source that appears on the system. Verifying sources usually calls for making repeated recordings over a long period of time. Using simple equipment and casual recording procedures, an unwary experimenter is likely to become overly excited by the appearance of some unexpected or unaccountable radio signal.

Terrestrial and man-made noise is an ever-present problem for radio astronomers; and if a casual observer interprets generated noise as a new kind of pulsar, he is likely to make a fool of himself.

Amateur radio astronomy is in a somewhat delicate phase of development, and it cannot afford to be discredited by overzealous experimenters who erroneously think they are receiving signals from extraterrestrial civilizations or vehicles.

DISCRETE-SOURCE PROJECTS

Discrete-source projects include some of the simplest formal projects an amateur radio astronomer can perform. Generally speaking, discrete-source projects deal with a single radio source that can be closely associated with an optical source. The experimenter records data from the object as it crosses his antenna beam or, if he has the appropriate kind of mounting, he can track it across the visible celestial hemisphere.

Discrete-source studies are especially appealing if the experimenter can correlate the position of the sources with optical objects viewed with the naked eye or with the help of an optical telescope. The most popular objects for discrete-source studies are Jupiter and its moons and the Sun.

One of the primary features of radio astronomy, however, is that the sources do not necessarily have to be those that are readily visible with the naked eye or small telescopes. Some of the most powerful radio sources in the universe are visible only through the world's largest optical instruments.

Solar Radio Studies

The Sun is by far the easiest object in the universe to study by means of radio astronomy. The source is obviously easy to find; and as long as the antenna and mount are suitably constructed, it is an easy object to track across the sky. The Sun is also noisy at every part of the radio spectrum. The regions of most intense noise and the quality of the noise, however, change from time to time.

The Sun radiates thermal radio energy as a result of the heat of fusion reaction, gives off intense bursts of nonthermal energy as the result of flare and sunspot activity, and it radiates a great deal of 21 cm *hydrogen line* energy. The Sun, in other words, is an ideal working model for all three sources of radio energy of current interest to radio astronomers.

An amateur radio astronomer has a moderate degree of control over the kind of radio energy he wants to observe. The key to separating the three kinds of energy is the fact that thermal radiation is most intense in the upper UHF range while nonthermal energy tends to be greater in the lower VHF range. Hydrogen-line radiation, of course, is tuned at 21 cm only.

It is relatively easy to eliminate all but nonthermal radiation by designing a radio telescope system that is tuned to the lower VHF range—in the commercial FM broadcast band, for instance. Eliminating most of the nonthermal energy and concentrating upon energy of thermal origin is a matter of assembling a system to operate in the upper UHF range.

It is impossible to eliminate all nonthermal energy, even in the microwave range. Tuning the system for one end of the useful radio spectrum does, however, give the experimenter some measure of selectivity; and it can lead to some unique experiments calling for time-scale comparisons between optical behavior such as sunspots and flares, and changes in thermal and nonthermal radiation.

An alternate technique for separating thermal from nonthermal solar radiation assumes the two kinds of radiation arrive at the Earth polarized differently. I know of no amateur work along this line, so it is up to some ambitious amateur to determine whether or not the two kinds of radiation are, indeed, polarized differently. (Hint: Compare signal levels from a series of antennas having vertical, horizontal, and helical polarization characteristics.)

Since the Sun is the most powerful radio source as observed from Earth, it has been the subject of intense scientific study almost from the beginning of radio technology itself. This does not mean the radio nature of the Sun is now completely understood. In fact, ongoing radio astronomy projects involve the Sun more than any other object in the sky.

There is very little known about the radio behavior of the Sun that is not at least partly clouded with some mysteries. The cyclic nature of sunspot acitivity, for example, has been studied for generations; but no one has yet posed a theory that accounts for it entirely. Most solar activity is less predictable than the peaks of sunspot activity, and an amateur radio astronomer can perhaps feel a dubious sense of belonging by knowing that the results of some of his solar radio projects might be no less puzzling to professional astronomers.

There is much data-gathering and theory-spinning to be done as far as the Sun is concerned. Professional astronomers have enlisted the aid of amateurs regarding the optical features of solar activity for a number of years. Working through the *Solar Division* of the American Association of Variable Star Observers (AAVSO), amateurs have supplied reams of valuable observational data that has eventually proved useful in the development of modern theories of the Sun. When amateur radio astronomy is finally established as a popular pursuit in the U.S., we can expect the professionals to solicit radio information about the Sun, too.

As far as amateur observers are concerned, solar radio studies can be classified as either *direct* or *indirect*. Studies of direct solar phenomena deal with events taking place directly on the Sun or in its atmosphere. Indirect studies deal with the influence solar activity exerts upon Earth's upper atmosphere.

Direct solar studies suitable for amateur investigators include correlating visual sunspot or flare activity with the intensity of radio bursts. It is clearly understood that sunspot activity creates intense bursts of nonthermal electromagnetic energy, but the correlation between sunspot activity and thermal radiation is not understood very well at all.

Indirect solar studies of special interest to radio amateurs are those that attempt to show a relationship between visual solar activity, thermal and nonthermal radio activity, and disturbances in terrestrial radio communications. There can be little doubt that solar magnetic storms play havoc with radio communications on the Earth by disrupting the structure of the Earth's upper D and E layers. Radio astronomy projects dealing with these phenomena can produce some satisfying results.

Amateurs with a special interest in solar radio astronomy should see the "Amateur Scientist" department of *Scientific American* magazine for September 1960. This article outlines plans for a longwave radio receiver and antenna suitable for detecting solar flares. The instructions are quite detailed and the procedures differ from any proposed in this book. Obsolescence of the semiconductors used in the receiver section has made them difficult to locate nowadays, but a knowledgeable experimenter should have no real trouble making suitable substitutions.

Nonthermal Studies of Jupiter

For the amateur radio astronomer, Jupiter is perhaps the most exciting and delightful radio source. The most intense nonthermal radiation from Jupiter is in the 18–30 MHz region, and the signals are noted for their peculiar audio sounds. To a chart recorder, radio signals from Jupiter often look like distant thunderstorm activity. Through a loudspeaker, Jovian broadcasts are frequently reproduced as seashore sounds or very distinctive chirps or staccato noises.

It can be very difficult for an enthusiastic amateur radio astronomer to convey some of his enthusiasm to others not especially interested in radio work: A set of squiggles on a graph? So what? But playing a tape recording of Jupiter's radio performances can spark the interest of any imaginative individual.

Thermal radiation from Jupiter is virtually nonexistent as far as the amateur is concerned. It is so slight that even in the upper UHF part of the spectrum the signals are barely distinguishable from internal radio noise. The causes of Jovian nonthermal radiation is poorly understood, but extensive radio studies carried out over a number of years make a strong case for a synchrotron theory. Apparently electrons in the upper atmosphere become trapped in a radiation belt similar to the Earth's Van Allen belt.

Some aspects of Jupiter's nonthermal radiation are predictable, but others are not. The signals vary greatly according to frequency, polarization, and duration. At one particular frequency and polarization, the radiation might be stronger than that of the Sun. A few magahertz away, the best radio installations can have trouble finding any Jovian signals at all.

As far as amateur radio astronomers are concerned, getting the most exciting kinds of signals from Jupiter is a matter of being in the right place at the right time. Beginners might do well to use the Sunday fisherman's approach: Set up a workable installation for one frequency and polarization, point the antenna in the general direction of Jupiter, and patiently wait for the planet to take the initiative.

The simplest radio project involving Jupiter entails the tape recordings of its unusual sounds. Using a fixed antenna with a beamwidth of 40° or so, the planet falls into the beam at least two hours out of every day. After making recordings for several months, the experimenter can build up a fairly good library of Jovian "concerts." Of course, the tapes can be edited to eliminate long periods of silence or recording sessions ruined by local interference.

An experimenter can add a bit of sophistication to the recording project by simultaneously recording WWV or CHU time signals on a second tape track. He can thus attempt to correlate the quality of Jupiter's sounds with its rotational period, the positions of its satellites, or any other parameter that seems to be of any real importance.

Taking a somewhat different route, an experimenter can make chart recordings of Jupiter's signals to obtain more objective evaluations of its signals. He can then attempt to correlate the recorded events with the rotational position of the planet and its moons. This kind of project can be especially exciting and meaningful if the experimenter takes photos or makes drawings of the planet as viewed through a telescope at the same time.

Sky and Telescope magazine, by the way, carries monthly prediction charts for the configuration of Jupiter's moons.

Amateurs graduating from the "Sunday fisherman" school of Jovian radio astronomy can attempt to pin down some of the more elusive features of the planet's signals. One can, for example, construct elaborate wideband antennas that enable him to scan the full spectrum of Jupiter's signals. While the planet is quiet at one frequency, it can be quite active at another.

It is also possible to use a pair of antennas and receivers to study the polarization characteristics of Jovian broadcasts. One beam antenna can be oriented vertically while the other is set horizontally; or if the experimenter chooses to use helical antennas, he can build one polarized clockwise and another polarized counterclockwise. In any case, the two receivers should be operated at the same frequency to get the most reliable and meaningful data for comparison.

The sophistication of Jovian radio projects is limited only by the experimenter's imagination and technical abilities. Automatic frequency scanning, audio time-constant and frequency analysis, and antenna tracking are all within the realm of possibility.

A special disadvantage of Jovian radio astronomy is the size of the antennas. Working in the decameter range, the reflector element on a beam antenna can be on the order of 25 feet long. And even allowing sloppy beamwidths of 40° or more, a 4- or 5-element beam stretches 25 feet or more into the sky. The antenna mounting scheme can be quite cumbersome, too—especially if it is rigged for tracking the planet.

A second disadvantage inherent in most radio work with Jupiter is the fact that its 18–30 MHz range of greatest activity is also crowded with industrial and amateur radio transmissions. Radio astronomy is, in principle, 24-hour astronomy; but most amateurs find that man-made radio interference makes daytime or business-hour observing in the decameter range all but impossible. From a practical standpoint, then, radio astronomy work with Jupiter is normally limited to the midnight-to-dawn hours and Sundays—times when the bands are relatively free of industrial and commercial interference.

The special advantage of working with Jupiter is that the receiver can be an ordinary shortwave receiver. The radio emissions from Jupiter are so intense that an ordinary receiver bandwidth of 10 kHz is normally adequate. (Most other radio astronomy projects require a bandwidth of 2 MHz or more.) If the sensitivity of the receiver is no better than 1 μV, however, the experimenter might find he needs an rf preamplifier.

Another advantage of building an installation for studying Jupiter is that it is equally suitable for studying the Sun's nonthermal radiation. With a single installation, then, an experimenter can carry out two entirely different kinds of projects at the same time.

Studies of Single Stars and Galaxies

Amateur radio astronomers have done little meaningful work with discrete sources outside the solar system. The work they have done has been largely limited to tracking a particular source across the sky, listening to its signals, and recording some data now and then.

There are perhaps a couple of reasons for this general lack of interest in discrete radio sources outside the solar system. For one thing, very few radio sources can be easily correlated with a particular optical object. Much of the fun in amateur radio astronomy is being able to see the source and listen to its signals at the same time. The other reason for general disinterest in remote sources is the fact that the sources, themselves, are rather uninteresting—their emissions seem to be nothing more than pure static that never changes in quality or amplitude. (Projects such as spotting pulsars is beyond the present capability of amateur equipment.)

So why bother studying the remote discrete sources? Whether he wants to or not, any amateur putting up a reasonably good radio telescope installation is going to deal with remote discrete sources. They are unavoidable. Sources such as Cygnus A and Cassiopeia A are so powerful that they are bound to influence observations at one time or another, regardless of the system's bandwidth and operating frequency.

And for the dedicated amateur, the thought that he is detecting radio signals generated at a time when his ancestors roamed the countryside wearing animal skins and carrying clubs may well be enough to sustain his interest.

There are some very compelling reasons for making a serious study of remote discrete sources. One of the hallmarks of radio astronomy—professional and amateur—is the element of surprise. The optical quality of many stars undergoes changes with time, and it is quite likely the same kind of thing happens to radio sources. Discovering a change in the quality of radio emissions from a source in space is a simple matter of looking at the right place at the right time. It is a simple matter in *principle* at least.

The odds against confirming a variable quality in a discrete radio source are tremendous. The odds that such changes take place at one time or another, however, are quite high. An amateur experimenter with the right kind of scientific attitude and patience can play the odds by studying single sources over a long period of time. Any careful, systematic, and on-the-ball scientific study of this sort is bound to turn up some positive results with time.

The equipment required for discrete-source studies of objects outside the solar system can be identical to that used for any kind of thermal or nonthermal radiation project. An amateur might spend a little bit of time each day running a routine check of known sources, then change over to projects that have more to offer in terms of excitement and positive results.

EXTENDED-SOURCE PROJECTS

An extended radio source can be defined as one made up of many discrete sources covering a relatively large segment of the sky. Sagittarius A (the galactic center) and the Milky Way are two examples of extended sources. These two, in fact, are the only extended radio sources readily observable with amateur equipment.

Sagittarius A is made up of millions upon millions of relatively weak suns, and any change in the radio signals from one of them cannot possibly influence the overall quality of radiation. Sagittarius A is thus a very stable radio source that will never show any gross changes on its own accord.

An amateur experimenter, however, can turn the stable quality of Satittarius A to his own use. Rather than expecting changes in its radio signals due to some internal phenomena, he can use it to test the radio transparency of the Earth's atmosphere.

The most interesting and fruitful extended-source project involves mapping the Milky Way at different wavelengths. Like Sagittarius A, the Milky Way is a powerful radio source only because it is made up of so many weak, discrete sources. Optically, the Milky Way appears as a delicate web of blue-white haze that meanders across the night sky. The meandering quality also appears on simple radio astronomy equipment; and the fact that the radio quality varies with frequency makes systematic mappings of the Milky Way especially interesting. The next section describes mapping techniques in great detail.

MAPPING PROJECTS

Mapping projects are by far the most ambitious projects the amateur radio astronomer can undertake. Many amateurs view mapping projects as the ultimate objective of all the preparations. Mapping procedures are time consuming and often frustrating; and unless the experimenter has access to a sophisticated computer system, the process can involve a great deal of pencil-and-paper calculations.

The whole idea of radio astronomy mapping is to generate a picture of the sky as it appears to a radio receiver at one particular frequency. (See Fig. 6-2.) In principle, making a map of the radio sky is much like mapping the sky as it appears to the naked eye. Radio maps, however, normally

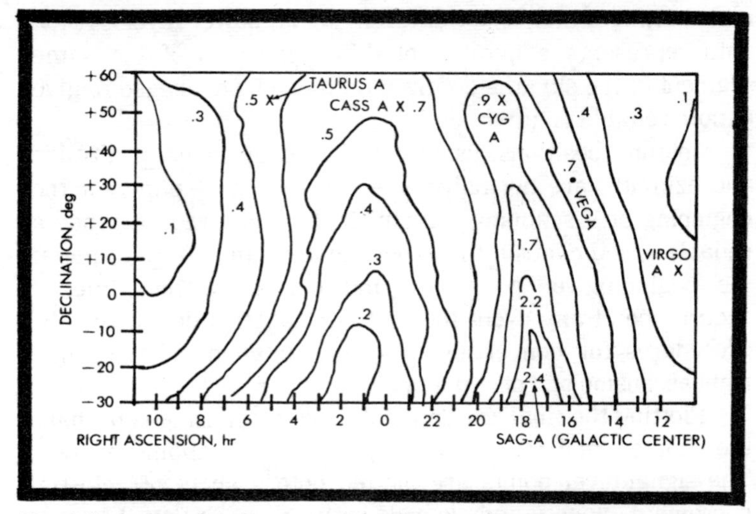

Fig. 6-2. A typical radio map of the sky. Gradient levels are in voltage units as read from the simple radio telescope system. The system could be calibrated to convert the level readings into standard flux units.

cannot show sources as discrete points—the lack of good radio resolution makes a single point source appear as a set of signal-level gradient lines that look something like altitude markings on a surveyor's terrestrial contour map.

The procedures for making a radio map of the sky can be broken down into four basic steps:

1. Gathering and recording of raw data
2. Digitizing the signal information
3. Converting the recorded times and antenna positions into right ascension and declination values
4. Plotting the digitized signal data onto a celestial map.

The information-gathering step in the map-making process concerns scanning the sky with the radio telescope system and recording the resulting signal levels on a chart recorder. Radio astronomers—even the professionals—are rarely certain the results of a single scan are perfectly reliable; so they scan the sky a number of times over a period of six months or more.

The purpose of the digitizing step is to convert the analog signal levels recorded on the chart paper into integers representing a number of different signal power levels. If the data represents a number of different scans of the same segment of the sky, the digitized data can be averaged to give a more reliable picture.

Antenna positions, usually reckoned in terms of altitude and azimuth, are normally noted on the chart paper at the beginning of a scanning session. If the chart recorder has a reliable speed control, the experimenter can note the time at the beginning of a session and determine the time of occurrence of any event thereafter. Chapter 4 deals with the procedures for converting time and antenna positions into right ascension and declination.

Plotting the finalized data onto a celestial map is perhaps the simplest and most rewarding part of a mapping project. The signal-level points are plotted onto a sheet of celestial coordinate paper and joined with lines drawn between equal-level points. Only when the project reaches this point does the dxperimenter know whether or not all his work is paying off.

Gathering Raw Data

The raw data required for generating star maps includes precise observation times, antenna positions, and the recorded signal from the radio telescope system. Since the radio antenna looks at only a small portion of the sky at any given instant, complete maps of the radio sky can be generated only by scanning the entire celestial sphere over some period of time. It is possible to scan the entire sky in a relatively short period of time by systematically moving the antenna to different positions. Scanning the sky by physically moving the antenna, however, places some special demands upon the antenna system and its mounting.

Rather than moving the antenna to scan the sky, most radio astronomers prefer to fix the antenna and let the Earth's rotation do the scanning for them. To simplify the coordinate conversion process, the experimenter should fix the antenna so that it always points to the southern leg of his meridian (180° azimuth). With the azimuth thus fixed, he can set the altitude to something like 10° and record data over an entire 24-hour period. If he is satisfied with the results, he can reset the altitude to something like 20° and gather raw data for another day. By increasing the antenna's altitude a certain number of degrees each day and allowing the azimuth to remain fixed, the experimenter eventually scans the entire celestial sphere.

Just how much the altitude setting should change after each scan depends upon the beamwidth of the antenna. For satisfactory coverage, the altitude should be increased no more than one-half the antenna's vertical beamwidth. If the antenna happens to have a vertical beamwidth of 20°, for example, the altitude increases should not exceed 10°.

Man-made radio noise, distant thunderstorms, and spurious noise generated within the electrical system can play havoc with the reliability of recorded data. Unless the experimenter is willing to babysit the system on a 24-hour basis and note all occurrences of unwanted signals, chances are quite good a burst of terrestrial noise could be heralded as an important discovery.

To reduce the possibility of plotting signals of terrestrial origin, radio astronomers should scan the same segment of sky at least four or five times. By comparing all the chart recordings and averaging the data, the experimenter can eliminate virtually all unwanted signals from the final results. Averaging a number of recordings of the same segment of sky, by the way, effectively increases the system's sensitivity as well.

Useful as the Sun is, it is also an ever-present nuisance. Whenever it passes through the antenna beam it plays havoc with the recordings and, more importantly, it *masks* signals from celestial radio sources behind it. About half the recordings for a complete mapping project must be made during daylight hours; and since it is important to ignore all tracings made during a solar transit, the experimenter has to develop a technique for filling in the "hole" it leaves in his map.

To get around the solar masking problem, the experimenter can spend several months recording data from the sky, ignoring signals from the region of the Sun. As described in Chapter 2, the Sun lags behind the motion of the celestial sphere by about 3 minutes each day. About two months after scanning all but the region of the sky around the Sun, then, the experimenter can scan those coordinates with fairly good results—the Sun will have dropped 30° behind the gap it left in the data before.

Jupiter also produces a masking effect that is further complicated by its annual performance of retrograde motion. The experimenter can eliminate signals from Jupiter using the same general procedures described in connection with the Sun. The only difference is that he must consult an almanac of *Sky and Telescope* magazine to determine the dates Jupiter is scheduled to move out of the way.

If a mapping project is to be carried out with any semblance of efficiency, the system must be operated for several months at a time on a 24-hour basis. This kind of operating schedule means the equipment has to operate unattended for very long periods of time. Ideally, the operator should only have to adjust the antenna altitude at the end of several scans, note the antenna setting and exact time at the beginning of a scan, change the chart paper now and then, and keep an eye on the system's overall performance.

The finest radio telescopes in the world barely approach the ideal, however, and any amateur is bound to encounter some frustrating bugs in the system. A radio telescope that is to be used for extensive mapping projects must be as reliable as possible, and that implies a thorough checkout of the system before launching a program. Equally important, it calls for occasional shutdowns for routine maintenance and calibration.

The fact that the system runs unattended most of the time also leads to a requirement for a circuit that automatically records the system's internal noise level. Every hour, a timing circuit of some sort should switch a fixed resistance in place of the antenna system just long enough to register the receiver's internal noise level on the chart recordings. And if the calibration timing circuit is accurate enough, it can also serve as a means of double-checking the recorder's time scale.

A complete data-gathering phase of a mapping project should produce raw signal data from the system, handwritten notes on the charts that indicate the antenna position time information, and calibration signals indicating the receiver's inherent noise level.

Digitizing the Signal Information

Unless an experimenter is fortunate enough to have access to an analog-to-digital converter, he has to digitize the raw analog signals on the charts by hand. All the work can be done on a prepared worksheet such as the one illustrated in Fig. 6-3.

The basic idea is to first measure off the raw signal levels at certain intervals of right ascension and declination. Just how fine the right ascension intervals should be depends upon the antenna's beamwidth (the resolution of the system). A good rule-of-thumb is to fix the right ascension intervals at $\beta/30$ hours, where β is the beamwidth of the antenna in degrees. If, for example, the antenna has an effective beamwidth of 15°, there is no point in scaling off the right ascension markings at any less than 15/30 or 0.5^h intervals.

The declination readings are completely dictated by the altitude setting of the antenna. Chapter 4 includes all the equations necessary for converting the time information into right ascension and antenna altitude into declination.

After scanning the chart recording in celestial coordinates along the time axis, the experimenter should measure off all the signal levels at the designated intervals and record them on the worksheet. He should then list the receiver noise figures closest to the designated intervals of right ascension.

The fourth column on the worksheet in Fig. 6-3 shows how the noise figures are subtracted from the raw data figures to produce a figure representing the actual signal level from the antenna. If a number of scans over the same region of right ascension and declination are to be averaged, the experimenter should use the final figures appearing in that fourth column.

Plotting the Final Data

The adjusted signal-level figures in the fourth column of the worksheet in Fig. 6-3 are the ones to be plotted on the celestial coordinate paper. The points are plotted according to

MAPPING DATA SHEET

Date 7-1-73
Alt. 40°

Time	Signal Level	Noise Calib	Adj Signal Level	Rounded Sig Level	RA	Dec
0140	1.27	.4	.87	.9	20.0	0
0210	1.16	.4	.76	.8	20.5	0
0240	1.14	.4	.74	.7	21.0	0
0310	00.93	.4	.53	.5	21.5	0
0340	00.83	.4	.42	.4	22.0	0
0410	00.77	.4	.37	.4	22.5	0
0440	00.71	.4	.31	.3	23.0	0
0510	00.68	.4	.28	.3	23.5	0
0540	00.51	.4	.19	.2	24.0	0

Fig. 6-3. Portion of a mapping worksheet. Data in the first three columns are taken directly from the recordings. The adjusted signal level is the difference between the recorded signal levels and the system's inherent noise level. Right ascension is computed from the known time of day, and the sidereal time in Greenwich (Appendix C). Declination uses the antenna's altitude setting—40° in this case—and the observer's geographical latitude.

their right ascension and declination, and each point should carry a figure representing the corresponding signal level.

Once the points are all plotted, the experimenter can connect the points having much the same signal-level values. Since the first maps are bound to include some discrepancies arising from gathering the raw data and manipulating the records, it is a good idea to make up some preliminary maps. The preliminary maps can be drawn up on ordinary typewriter paper. The finished map—the result of the most ambitious and demanding project an amateur radio astronomer can perform—can be as large and dramatic as desired.

Summary of Mapping Procedures

1. Set the azimuth of the antenna to 180°—due south.
2. Set the antenna altitude to 0°—toward the true horizon.
3. Record the signal and time data for 24 hours, changing tapes or chart paper and checking noise figures as necessary.
4. Set the antenna altitude to 10, 20, 30, 40, 50, 60, 70, 80, and 90 degrees, repeating step 3 between settings.

5. If the recordings are made on magnetic tape, play back the recordings in the order they were made, monitoring the raw signal output with a voltmeter. If the recordings were made on chart paper, search through them for signal responses of the type described in step 6.
6. Wherever the output signal seems to increase a significant amount, study that section of the tape or chart carefully. Manually record the signal levels and time reports at 30-minute intervals throughout the signal response.
7. Transform the tabulated time, azimuth, and altitude information into right ascension and declination (see Chapter 4).
8. Round off the tabulated raw data to about 10 different numbers.
9. Prepare a celestial coordinate chart, and place a few bright optical objects onto it for reference purposes.
10. Plot the right ascension and declination of all tabulated signal levels.
11. Connect the plotted points having like signal-level values.

Equipment and Project Planning

Professional radio astronomy has only recently passed the point in its development where serious and highly productive work can begin. For the most part, electrical engineers have guided the development of this new technology. The pioneers of radio astronomy were all electrical engineers; and, generally speaking, today's leading radio astronomers are trained electrical engineers who have absorbed an interest in astronomy. Furthermore, most university radio telescope installations are still funded and operated by departments of electrical engineering rather than departments of physics or astronomy.

It is no coincidence that electrical engineers have been the backbone of radio astronomy to date. Radio astronomy combines the theories and know-how of two vastly different specialities; and only after the electronic systems have been perfected can an astronomer with little electronics experience enter the picture in any meaningful way.

Amateur radio astronomy is now in its infancy. There is, in fact, very little to be said about its progress simply because there has been so little progress of notable significance. Radio astronomy is bound to become a serious hobby for amateur experimenters in the years to come; but as in the case of professional radio astronomy, it will be up to electronically oriented amateurs to lead the way.

At the present time, an amateur radio astronomer must know electronics. Some knowledge of astronomy is important, too; but without mastering some electronics skills, there is no hope of accomplishing anything that is remotely meaningful or interesting. The main problem is that there are no off-the-shelf, ready-to-go instruments specifically designed for amateur radio astronomy. If an amateur astronomer could buy a radio telescope installation (or even a kit) from an electronics shop, the situation would be quite different. As it stands now, however, every amateur radio telescope installation is a peculiar combination of standard and customized circuits that are jury-rigged together to perform what is, hopefully, some useful and interesting radio astronomy work.

Assembling an amateur radio telescope installation is pretty much a matter of trading off the basic requirements for any radio telescope system against the equipment available to do the job. Every amateur installation is essentially a custom system, and this chapter deals with the facts and techniques an amateur must consider when planning his own system. The chapter is particularly relevant to amateurs who do not have the know-how or inclination to work with the formal design equations in Chapter 5.

GENERAL REQUIREMENTS

From a systems point of view, the essential parts of a radio telescope include antenna, receiver, and some sort of readout device. From an operations point of view, the list of devices must be extended to include some means of calibrating the system, recording and interpreting data, and maneuvering the antenna. At first glance, it might not appear very difficult to assemble a complete radio telescope installation: Simply build an antenna, connect it to a receiver and recorder, and start collecting data. It is not all that simple, however. Assembling a useful radio telescope involves making tradeoffs all along the way—every decision about one part of the system exerts a powerful influence upon the requirements for other subsystems.

Electrical Requirements

The electrical requirements are by far the most important. An ambitious experimenter can build a radio telescope from the most elaborate and expensive kinds of radio equipment available today; but if any one part fails to meet the basic requirements for a working system, the user is bound to be disappointed with the results. On the other hand, an experimenter can put together a workable radio telescope from standard electronic devices and parts—FM receivers, a spare voltmeter, plumbing supplies purchased from the local hardware store, etc. It's all a matter of meeting the specifications. The sophistication of the equipment is of secondary importance.

Ideally, a radio telescope system has high gain, wide bandwidth, low inherent noise, and high antenna directivity. The first three requirements are almost self-contradictory. The fourth requirement, high antenna directivity, complements the need for high power gain, but it tends to run counter to the need for wide bandwidth. These are the most basic requirements for a radio telescope system, and all other design considerations must give way to them. The fact that these electrical requirements are almost wholly contradictory leads to the real challenge of amateur radio telescope design.

Table 7-1 lists some minimal electrical specifications for amateur radio telescope receivers and antennas. These are only ballpark figures, but they can serve as guides to selecting equipment and designing the antenna. Besides, it is often difficult to determine the actual operating specifications of a receiver and antenna, especially when they are used in an abnormal configuration such as a radio telescope. The best a beginner can do if he cannot handle the equations in Chapter 5

Table 7-1. Basic System Specifications

Parameter	Minimum	Normal	Excellent
Antenna gain	10 dB	15 dB	20 dB
Receiver sensitivity	5 μV	1 μV	0.5 μV
Receiver bandwidth	100 kHz	2 MHz	6 MHz

The **Minimum** system is capable of detecting signals from Cassiopeia A and the galactic plane. The **Normal** and **Excellent** are capable of working at least five sources outside the solar system.

is to select equipment that meets the specifications in a very general way, try it out, and make any necessary modifications to get the system working to his own satisfaction.

It must be noted at this point that the basic requirements for a workable radio telescope system are met only by the proper selection of both receiver and antenna. It is rather pointless to argue whether the receiver or antenna is more important. The basic requirements push both elements to their technological limits. And it is virtually impossible to find a "better" version of one that makes up for any shortcomings in the other.

Although receiver and antenna stand on the same level as far as system performance is concerned, professional and experienced amateur radio astronomers tend to stress the importance of the antenna. This emphasis upon antenna performance comes about because ready-made receivers of suitable quality are available from many different sources. Antennas, on the other hand, must be designed, built, and tested from scratch. Unless an experimenter chooses to build his own receiver, then, most of his time, patience, and technical know-how will go into the construction of the antenna system. Therein lies the natural tendency to emphasize the role of antennas in radio astronomy.

The nature and quality of the readout device plays an important role in meeting the overall electrical specifications, too. Readout devices, whether they are as simple as a voltmeter or as complex as a chart recorder, are subject to limitations of gain, bandwidth, and noise. Relying upon a readout device to make up for any shortcomings in the receiver or antenna merely compounds the difficult problem of trading off all the contradictory electrical requirements for the system.

Practical Considerations

A host of practical considerations enter the picture once an experimenter meets the basic electrical requirements for his radio telescope system. Availability of the equipment, of course, is the primary practical consideration in the assembly of a working system. A second kind of practical problem concerns the location of the installation itself.

The question of equipment availability usually boils down to one of dollars and cents. An experimenter who wants to assemble a radio telescope from all-new equipment can expect to invest at least $100 before getting the system on the air. Most of the working systems described in this book, however, can be assembled from the kinds of equipment radio amateurs and electronics technicians already have at hand. In such instances, the only dollar outlay is for the antenna system—an amount that can vary anywhere between $ 5 and $ 50.

There are three facets to the problem of locating the installation: (1) The space must be large enough to set up and maneuver the antenna, (2) the experimenter must have an unobstructed view of the portion of the sky he wants to study, and (3) the area must be relatively free of man-made radio noise such as that from diathermy machines, industrial power controls, and extremely heavy auto traffic.

Table 7-2. Summary of Antennas and their Characteristics

Antenna Type	Advantages	Disadvantages
Multielement beam (Chapter 8)	High gain Relatively simple construction Offers a variety of impedance matching techniques	Narrow bandwidth Polarized in one plane
Beam array	High gain Any linear polarization possible Fairly wideband design possible Many impedance matching options	Mechanically awkward Requires a great deal of working space
Single helix (Chapter 9)	Inherently wideband Circular polarization Moderate gain	Mechanically awkward at longer wavelengths Polarized in one direction
Helical array	High gain Linear or circular polarization possible Inherently wideband	Mechanically awkward at some frequencies
Parabolic reflector	Extremely directive High gain	Extremely difficult to build

Table 7-3. Summary of Antenna Mountings and Their Characteristics

Antenna Mounting Type	Advantages	Disadvantages
Fixed (Chapter 10)	Simple construction	Not steerable in any plane
Altitude (Chapter 8)	Relatively simple construction	Azimuth is fixed
Altiazimuth	Fully steerable	Relatively complicated construction Awkward at decameter wavelengths Works in horizon coordinate system
Equatorial	Fully steerable Works in celestial coordinate systems	Complicated construction

All of these electrical and practical considerations are described in more detail later in this chapter.

PLANNING AN INSTALLATION

Every radio telescope, whether constructed by professionals or amateurs, represents the result of a whole series of technical and practical tradeoffs. Every working system is a compromise between what the experimenter wanted to do in the first place and the equipment available to him in the end.

It is only through careful and realistic planning that a completed system can approach the experimenter's ideal. The remainder of this chapter deals with the kinds of projects and equipment that seem most attractive to amateur radio astronomers at the present time. Tables 7-2 through 7-8 summarize the main points presented in this chapter.

The discussions and tables will be of most help to those individuals who are launching their first radio astronomy projects. Hopefully, more experienced and sophisticated amateurs will find the tables to be a source of ideas for building new systems or expanding the capabilities of existing ones.

Selecting a Project

Chapter 6 describes the kinds of projects that are especially suitable for amateur radio astronomers. Every project requires a slightly different inventory of equipment, and an experimenter can make real headway in his equipment selection only after defining what he wants to do with it.

For planning purposes, radio astronomy projects fall into three general categories: studying nonthermal sources in the solar system, mapping the radio sky, and locating and studying discrete and extended sources outside the solar system.

The selection of a project has a direct bearing upon the system's operating frequency, type of antenna mounting, the system's gain and resolution, and the quality of the readout and recording sections (Tables 7-2, 7-3, and 7-4). Of course, the selection of an operating frequency has a direct bearing upon the receiver specifications; the type of antenna mounting influences the choice of antenna size; and so on. An experimenter can indeed assemble a system and decide what

Table 7-4. Summary of Receiver Options

Receiver Type	Advantages	Disadvantages
Shortwave communications	High gain Suitable converter and preamp circuits available	Narrow bandwidth
Commercial FM receiver (Chapter 8)	Good gain Wideband characteristics Low noise	Restricted to 88 — 108 MHz operating frequency
Special homemade receiver (Chapter 9)	High gain Wide bandwidth Low noise	Requires know-how to build and properly align
Custom receiver made from commercially available modules	Best possible gain, bandwidth, and noise figures	Expensive

Table 7-5. Summary of Record—Readout Devices

Record/Readout Device	Advantages	Disadvantages
Voltmeter	Readily available	Requires constant attention during a recording session
Chart recorder	Requires little or no attention during a recording session Provides permanent records	Initially expensive
Tape recorder	Requires little attention during a recording session Provides permanent records Can be used for other purposes	Data must be transcribed at a later time

he wants to do with it later on; but that brand of turned-around planning greatly reduces the chances of getting any interesting and meaningful results.

Selecting an Operating Frequency

The radio astronomy frequency spectrum extends from about 18 MHz to 30 GHz or so. There is little point in trying to work outside this range because there is so very little thermal radiation below the lower limit and because the Earth's atmosphere is opaque to frequencies beyond the upper bound. It isn't merely the lack of a suitable technology that places these limits upon radio astronomy frequencies—it is the very nature of radio emission and propagation.

Practical considerations narrow the useful range of operating frequencies even more. A few amateurs might try their hand at modifying some 1296 MHz amateur radio equipment for microwave projects; but for the most part, amateurs find working with centimeter wavelengths to be too demanding in terms of know-how and patience.

The UHF bands are wide open to amateur experimenters; but unfortunately receiving equipment with the required noise figures is not readily available. Anyone wanting to work in the UHF range should thus be prepared to construct his own receiver.

Table 7-6. Equipment Options for Nonthermal Solar-System Studies

Readout-recorder options:	voltmeter chart recorder tape recorder system
Special difficulties:	1. Antenna requires a great deal of space 2. System susceptible to man-made interference
Operating frequency:	18 – 60 MHz
Receiver options:	shortwave receiver homemade receiver
Antenna options:	5-element beam beam array corner reflector
Antenna mountings:	fixed

At the low end of the radio astronomy spectrum, only two kinds of projects are of any real significance: studying nonthermal radiation from Jupiter and recording solar flare and sunspot activity. Any other kind of radio astronomy work demands an antenna with much higher gain. It is not altogether impossible to build an antenna for 18 MHz that has a gain on the order of 20 dB, but it would be an enormously ambitious undertaking.

For practical purposes, then, general amateur radio astronomy work (excluding nonthermal studies in the solar

Table 7-7. Equipment Options for Detecting Galactic Sources

Operating frequency:	60 – 180 MHz
Receiver options:	commercial FM receiver homemade wideband receiver
Antenna options:	13-element beam helical array
Antenna mountings:	any
Readout-recorder options:	any
Special difficulties:	Difficulties are dictated mainly by the characteristics of the selected antenna mounting and readout – record device.

system) falls into the 100–400 MHz part of the radio spectrum—loosely, the VHF band. This is not to say an ambitious and accomplished amateur cannot perform some interesting and useful experiments outside this range. At this point in the development of amateur radio astronomy, though, it is more desirable to stay within a frequency range that promises the best results with the least amount of effort and expense.

VHF technology is fairly well established today. Receivers with the required gain and bandwidth are readily available, and radio amateurs are finding the prospects of designing and constructing VHF equipment more attractive. It is, in fact, the firm establishment of this relatively new technology that makes amateur radio astronomy possible today.

Once an experimenter's thinking is oriented toward the VHF bands, he is in a position to select a receiver system. The receiver, of course, restricts the operating frequency of the system to a much narrower band; and since the VHF bands are normally crowded with man-made transmissions of all sorts, the amateur finds his working frequencies limited to quiet places between commercial FM, TV, amateur bands, and other private and public service bands.

Selecting a Receiver

The receiver portion of a radio telescope installation can be a completely self-contained unit such as a commercial FM radio or a combination of sections such as a preamplifier, tuner, and audio amplifier. In any case, the choice of a receiver is often a matter of trading off what is readily available against the basic electrical requirements for a working system.

Few amateurs have any difficulty rounding up a commercial FM receiver. Providing the unit is of moderately good quality and in good working order, it can serve quite well in a beginner's radio telescope installation. What's more, it is possible to tap off the appropriate signals without performing serious surgery on the circuitry—the set can thus be restored to its original application after it has served its special function in a radio telescope setup. Some amateurs, in fact, arrange their system so they can steal the family FM radio just long enough to perform some experiments now and then.

Any good FM radio has the necessary bandwidth for radio astronomy applications. In some instances, however, the units might have a rather poor signal-to-noise ratio. If in-line tests show signals from the Sun barely reach out of the receiver's inherent noise, it is necessary to attach the appropriate kind of rf preamplifier ahead of the receiver. FM radio and TV preamplifiers of suitable quality are available through electronics shops and catalogs for about $ 25.

Shortwave and amateur-band receivers can be used for certain kinds of radio astronomy projects; namely, those concerning the most powerful radio sources, such as the Sun, Jupiter, Sagittarius A, and the Milky Way. The problem is not with the sensitivity of such receivers—the sensitivity of a good amateur receiver is 10 to 100 times better than some TV sets display. The real trouble is that shortwave and amateur-band receivers have rf bandwidths on the order of 10 kHz or less; and as described in Chapter 5, a wide predetection bandwidth is essential to the overall sensitivity of a radio telescope.

Ambitious amateurs can construct receivers from scratch, keeping in mind the fact that the circuit must have qualities of high gain, wide bandwidth, and low noise. These specifications are difficult to meet, and there are too few circuits published in electronics books and magazines that can do the job. Anyone wanting to do his own engineering might use some of the techniques outlined for the homemade receiver system described in Chapter 9.

A convenient but expensive compromise between using ready-made receivers and building one from scratch involves modular rf and i-f sections now available from a number of manufacturers. These radio sections are available on a quasi-custom basis, and they can be selected to operate at just about any desired frequency. All the experimenter has to do is devise a stable dc power supply and interconnect the modules. Consult an industrial electronics catalog, such as the *Electronic Engineers Master* (EEM), for company names, specifications, prices, and availability.

Selecting an Antenna

The selection of an operating frequency dictates the general dimensions of the antenna system, while the kinds of projects planned for the system have a direct bearing on the required antenna mounting. There are dozens of basic types of antennas that could, in principle, serve in a radio telescope system, but the necessary dimensions and desired mounting configuration go a long way toward narrowing the field of possible antenna types to two or three.

Before describing the two types of antenna designs that are presently the most suitable for amateur radio astronomy work, it is important to note the undesirability of parabolic reflectors. Giant parabolic reflectors have become the hallmark of professional radio astronomy over the years—virtually all major installations use them in one form or another. As far as amateur radio astronomy is concerned however, parabolic reflectors are wholly unsuitable.

The most devastating disadvantage of parabolic reflectors is their inherent inefficiency at *milliwave* frequencies—those wavelengths for which it is not practical for the amateur to design or operate a parabolic antenna. Below the gigahertz range these antennas become so large and difficult to maneuver that only professionals can afford the machinery needed to steer them. At the center of the useful amateur radio spectrum, parabolic reflectors are less than 70% efficient in terms of gain or directivity. To recoup the losses, the reflector must be made at least 30% larger than elementary design calculations show.

From a practical standpoint, parabolic antennas are about the worst possible choice among all conventional antenna types. Constructing a precision parabola that is more than 5 ft in diameter is, at best, an ambitious, time consuming, expensive, and frustrating undertaking. Even after it is built, its precision drops off as normal movements and stresses warp the framework. Parabolics are inefficient in the first place, and the normal physical stresses compound the trouble.

Unless an amateur experimenter is more interested in staging a good show for onlookers than he is in getting good results from his installation, it is better to avoid the thought of using parabolic reflector antennas.

Knowledgeable amateurs recommend either *beam* or *helical* antennas for amateur radio astronomy work. Furthermore, they suggest using long beam antennas for working the VHF part of the spectrum and helical bays for the UHF work.

Properly matched beam antennas with at least 12 elements meet the basic requirements for amateur radio telescopes. Using 12 elements spaced about one-half wavelength apart produces a beamwidth of about 20° and a gain on the order of 20 dB. It is possible to increase both the resolution and gain by adding more elements or resorting to a bay of several smaller ones.

Unfortunately, beam antennas produce their peak gains at their resonant frequency. The gain drops off rather sharply on either side of resonance, making truly wideband operation rather difficult. Resonant beams nevertheless work tolerably well in conjunction with a wideband receiver.

One of the primary advantages of a helical antenna is its wide bandwidth. Most helical antennas can operate satisfactorily over a range of 200 to 300 MHz, making them obvious choices for radio astronomy work. Helicals, however, tend to be something of a mechanical nuisance at VHF; and amateur radio astronomers normally apply them only to UHF work.

Selecting an Antenna Mounting

Antenna mountings of interest to amateur radio astronomers are those that are completely fixed, those adjustable in altitude only, and those adjustable in both altitude and azimuth. If it weren't for some severe practical limitations, radio astronomers would probably select the altiazimuth (altitude–azimuth) mount—or the equatorial variation of it—in every case. The larger the antenna system, however, the more difficult it becomes to adjust and lock its position; so the most desirable kinds of altiazimuth mount give way to altitude mount in the VHF range; and the altitude mount give way to fixed mounts in the decameter range.

It should be noted that the selection of a mounting system is based upon purely practical considerations. There is no reason why a massive 20 MHz antenna system could not be attached to an equally elaborate equatorial or altiazimuth mount. Many of the giant dish antennas in professional installations are fully steerable in altitude *and* azimuth. Anyone with the time, know-how, and ability to acquire the necessary materials can indeed assemble an altiazimuth mounting for any antenna. In keeping with the more down-to-earth spirit of this book, however, altiazimuth mounts are recommended only where they are most practical and useful.

Because of their tremendous size and generally awkward construction, antennas used for nonthermal studies in the solar system are usually mounted in a fixed position. Such mountings are normally constructed on the spot, and aimed at the point in the sky where the ecliptic crosses the local meridian. The mounting can be aimed due south with the help of a simple magnetic compass, and the altitude can be fixed in the direction of the ecliptic with a drawing compass and plumb bob.

There are two main reasons for aiming a fixed antenna to the point where the ecliptic crosses the meridian. For one thing, Jupiter and the Sun both travel through the ecliptic; and by shooting for the ecliptic with a fixed antenna system, the experimenter can expect to observe both objects once each day. Then, too, aligning the antenna on the meridian greatly simplifies the mathematical procedures for determining the times the objects are scheduled to cross the beam. The antenna system described in Chapter 10 is fixed in this manner.

An *altitude* mount can be used for moderately long beam antennas cut for the lower VHF range. The antenna in this instance is somewhat more maneuverable than a 20 MHz version, but it is usually too cumbersome for full altiazimuth motion. An altitude mount thus represents a practical compromise between a totally fixed mount and one that is free to move in all directions.

As in the case of a fixed mount, altitude mounts are normally set with their azimuth fixed on the meridian. The user can thus pick up signals from sources crossing the meridian with altitudes anywhere between 0° and 90°. Any celestial body that is ever in the sky crosses the meridian at one time or another during a 24-hour period. If an altitude mounting is arranged so the user can swing it from the southern horizon, through the overhead point, and to the northern horizon, he has the capability of studying any radio source he chooses. The only disadvantage is that he has to wait for the source to cross his meridian.

The only reason for fixing an altitude mounting so that it always points the antenna toward the meridian is that it simplifies the calculations for predicting transit times and identifying sources. Altitude mounts are suitable for VHF mapping projects; and if the data is always coming from the meridian, the mapping task is greatly simplified.

An altitude mount requires some schemes for setting the altitude of the antenna and fixing it at some desired point. The technique described for the 110 MHz system in Chapter 8 can serve as a starting point for devising some useful altitude adjustment schemes.

Altiazimuth mounts are fully steerable in both altitude and azimuth, and their primary advantage is that the user does not have to wait for a source to cross the meridian before he can observe it. Any source that is above the horizon is a good target for an altiazimuth mount; and because of this particular feature, it is especially appealing to beginners and those wanting to collect data from a number of sources in a very short period of time.

Altiazimuth mounts, however, are highly impractical for all but the smallest antenna systems. Even if an experimenter happens to have the space for maneuvering a large antenna system on an altiazimuth mount, he will find mechanical stresses giving him some problems.

An altiazimuth mount requires two sets of scales: one labeled from 0° through 90° for altitude settings and another labeled from 0° through 360° for azimuth settings. The altitude scale must be carefully aligned with a drawing compass and plumb bob, and the azimuth scale has to be adjusted with the help of a magnetic compass.

One other difficulty associated with an altiazimuth mount is the complexity of the mathematics for converting the coordinates to and from the horizon and celestial systems. While these calculations are quite simple when working the meridian, they become rather complex for off-meridian work. An equatorial mount, a variation of the basic altiazimuth scheme, completely eliminates the coordinate-conversion problem.

An equatorial mount is the same as an altiazimuth mount in principle. The only real difference is that the azimuth ring is tilted so that its axis points directly to Polaris, or declination 0°. The two scales are labeled in terms of declination and hour angles rather than altitude and azimuth. With such a system, the user works entirely in the celestial coordinate system, and he only has to calculate right ascension from the sidereal time and the hour-angle reading.

Selecting a Readout–Recording System

The whole purpose of the readout device in a radio telescope is to convert electrical signals from the receiver into a form more meaningful to human operators. A good readout scheme must also include some means of recording the data for later study.

An experimenter could, of course, attempt to judge changes in signal level by listening to the static from a loudspeaker, but that is not a very objective technique. A useful radio telescope readout–recording system can be as simple as a voltmeter and a sheet of paper for recording voltage levels, or as elaborate as a multispeed oscillograph. As in the case of the main receiver system, the selection of a readout–recording system is pretty much determined by the equipment the experimenter has available to him.

The simplest possible kind of readout–recording system consists of an ordinary electronic voltmeter connected directly to the audio output of the receiver. The user can record voltage levels on a sheet of paper at regular intervals and later plot the readings on a graph.

The main advantage of a voltmeter readout is its availability—most radio amateurs and electronics technicians have ready access to a spare meter. Another advantage is that there is no need to make any serious modifications to the receiver circuit. The voltmeter leads can be connected directly to the speaker leads in the receiver, and quickly disconnected when the radio or voltmeter has to be used elsewhere.

The primary disadvantage of a voltmeter readout scheme is that the user must be on hand to record the data, making it virtually impossible to conduct extensive mapping projects. It is possible to overcome this particular handicap, however, by recording the signals on magnetic tape.

A magnetic tape readout—recording scheme involves two data-gathering steps: recording the raw information from the receiver onto the tape and playing back the recording through a voltmeter at some convenient time. A stereo tape recorder is especially useful in this instance because the experimenter can record the raw signal data on one track and time signals from WWV or CHU on the other. In this manner, he has an accurate and reliable correlation between changes in signal level and the time the changes occur.

The problems associated with the tape recorder and voltmeter schemes are that the experimenter still has to record data by hand and he has to account for audio filtering and distortion introduced by the tape system.

The most direct, convenient, and reliable way to record signals from a radio telescope is by means of an *oscillograph*, or paper chart recorder. Chart recorders allow the system to be operated day in and day out without any special human intervention; and when a recording session is completed, the experimenter can readily detect and interpret interesting peaks and valleys in the results.

The only disadvantage of a chart recording system is its initial cost. Chart recorders, even those of only moderate quality, cost in excess of $100. The author's experience, however, shows that there are a number of used chart recorders lying around in industrial and school laboratories. A serious amateur who is willing to do some legwork can usually round up a suitable chart recorder for a modest price or on a long-term loan basis.

Using an old chart recorder brings up the possibility of a problem that can render the whole scheme useless—the availability of fresh recording paper. Before buying an old chart recorder, the experimenter must make certain he can get paper that fits it.

Chart recorders are available with a number of different paper speeds. The chart speed required for radio astronomy work depends largely upon the kinds of projects being carried out. For long-term mapping projects, a chart speed on the order of 5 mm per hour is most appropriate. For on-the-spot recordings of transient signals—nonthermal Jovian bursts, for instance—a chart speed of about 10 mm per minute is best.

A Simple 110 MHz Radio Telescope

The low-cost 110 MHz radio telescope system described in this chapter was designed for two principal reasons: to test the mechanical performance of a very long beam antenna and to check the usefulness of a simple data recording scheme. The system illustrated in Fig. 8-1 performed quite well—surprisingly well, in fact, considering its simplicity and low cost. Beginning radio astronomers, especially those who do not have ready access to much amateur radio equipment, should find this drift radiometer an attractive initial project.

EQUIPMENT AND BASIC PROCEDURES

The antenna for this system is a 13-element yagi cut to about 110 MHz. The receiver is a conventional AM–FM

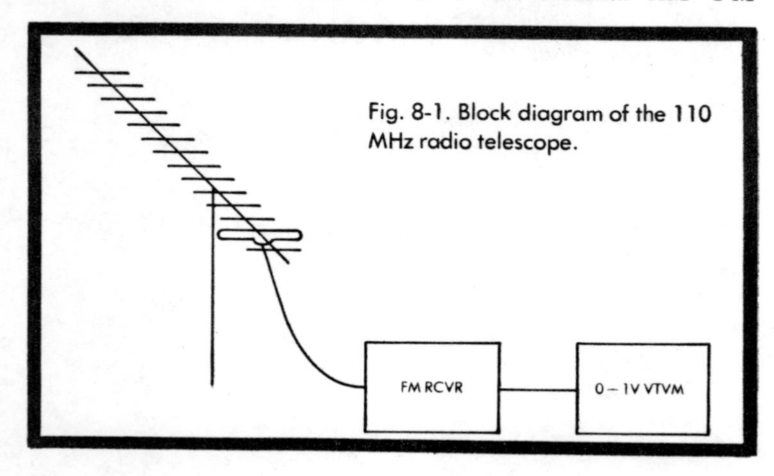

Fig. 8-1. Block diagram of the 110 MHz radio telescope.

broadcast radio of the department store variety, and the readout device is a VTVM that is connected directly to the speaker leads in the radio. The antenna and its mount cost about $ 30 and requires one weekend to assemble and erect.

Beam antennas have narrow bandwidths—a feature that, in theory at least, makes them unattractive for radio astronomy work. Considering the clumsiness of a 110 MHz helical array of comparable gain, however, suffering a loss of bandwidth seems to be the lesser evil. As shown in Fig. 8-3, the beam is a bit longer than 30 feet. The mast is 16 feet high to allow the beam to be rotated to the zenith. The whole structure is rather large and cumbersome, so I used an altitude mount with the azimuth fixed on the meridian.

Matching the low impedance of the 13-element yagi to the transmission line and receiver presented something of a problem in theory, but the actual as-built system works quite well by using a quarter-wave balun at the antenna and running a 50-foot length of 50Ω coax to the receiver. An engineering purist might shudder at the thought of a 52Ω–300Ω mismatch at the receiver terminals—but it works. A better match would improve the system's performance, of course.

In keeping with the demand for simplicity and low cost, the coax is connected to the antenna terminals on the receiver by means of an ordinary TV antenna clip. Getting a noise figure is thus a simple matter of removing the clip and replacing the antenna connection with a 50Ω resistor.

The AM–FM radio is, of course, operated in the FM mode and at the extreme high end of the band. One of the primary reasons for selecting 110 MHz as an operating frequency in this case is that it rests between the commercial FM band and the marine-aviation bands. The 110 MHz part of the spectrum, in other words, is relatively free of terrestrial radio transmissions. Most FM radios are designed to tune several hundred megahertz above and below the FM band (88 to 108 MHz), so an experimenter should have no trouble tuning 110 MHz.

The radio is always operated with the tone control set for maximum treble response and the AFC switch turned off. When recording the system's noise figure or taking antenna readings, the volume control is set for full audio gain.

Tests with an audio-modulated 110 MHz signal from an rf signal generator showed that the FM radio could faithfully reproduce an output voltage as high as 1.1V. A VTVM with a 1.5V ac scale thus made a suitable readout device.

During a normal recording session, the radio is set to full audio gain; the antenna is disconnected from it and replaced with a 50Ω resistor. The reading on the VTVM under these conditions represents the system's inherent noise level, and the figure is recorded so that it can be subtracted from the data at some later time.

To make the subtraction-of-noise procedure as simple as possible, the zero-adjust knob on the VTVM can be set to show some convenient low reading on the scale. It would have been nice if the VTVM could have been adjusted to 0V with a fixed resistor across the receiver's input terminals, but the zero-adjust range on my VTVM could not compensate for all the noise. It was thus necessary to consider a reading of 700 mV as a "signal zero" level.

After "zeroing out" the normal radio noise, the 50Ω resistor is removed from the radio's antenna terminals and replaced with transmission line of the same characteristic impedance. The increase in voltage that occurs at this point represents the amount of signal power at the antenna, less the losses due to impedance mismatches. Since the radio's inherent noise level and the output of the VTVM circuit tend to drift with changing line voltage and operating temperature, it is a good idea to zero the noise level before every antenna reading.

SOME PERFORMANCE SPECIFICATIONS

As expected, the 13-element yagi exhibited a beamwidth of about 30°. Considering the simplicity and low cost of the system, 30° is an acceptable resolving capability. It is, in fact, adequate for resolving all the major radio sources.

The background noise encountered on the completed system varied between 50 and 70 mV above internal noise.

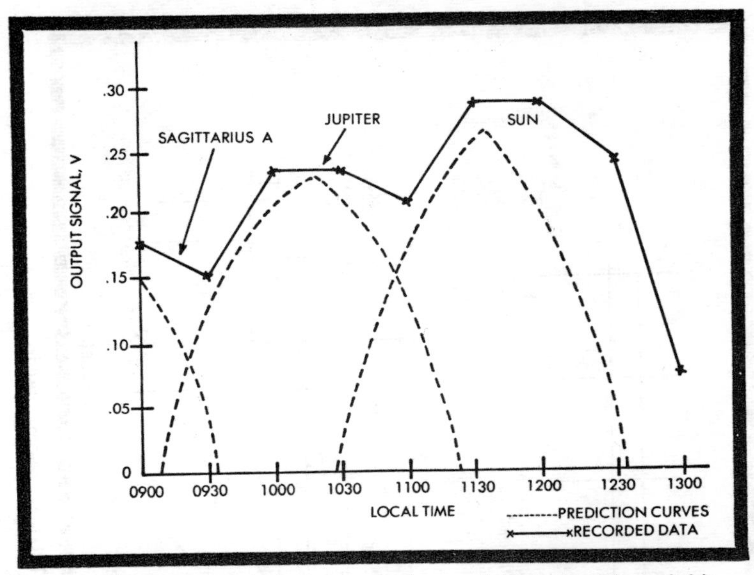

Fig. 8-2. Signal response to a series of powerful sources. The solid lines represent data recorded from the 110 MHz telescope. The broken lines are prediction curves based upon a knowledge of the events taking place on the celestial meridian that morning.

Jupiter and the galactic plane in the region of Monoceros showed peak signal levels of 150 mV. Sagittarius A registered a peak signal of 200 mV, while the Sun drove the amplifiers into saturation at about 270 mV of signal.

Some unidentified radio transmissions occasionally interfered with the readings, but on the whole the system was relatively free of man-made interference. The only consistent source of electrical disturbance was due to a neighbor's passion for revving a '37 Plymouth.

With an antenna resolution of 30°, it is pointless to take antenna readings more than once every half hour. I made a number of readings at half-hour intervals with the antenna as selected radio sources crossed the antenna beam. Figure 8-2 compares the system's response to a set of prediction curves.

ANTENNA CONSTRUCTION AND MOUNTING DETAILS

One of the basic features of the antenna described in this section is its modular quality. The whole antenna, in other

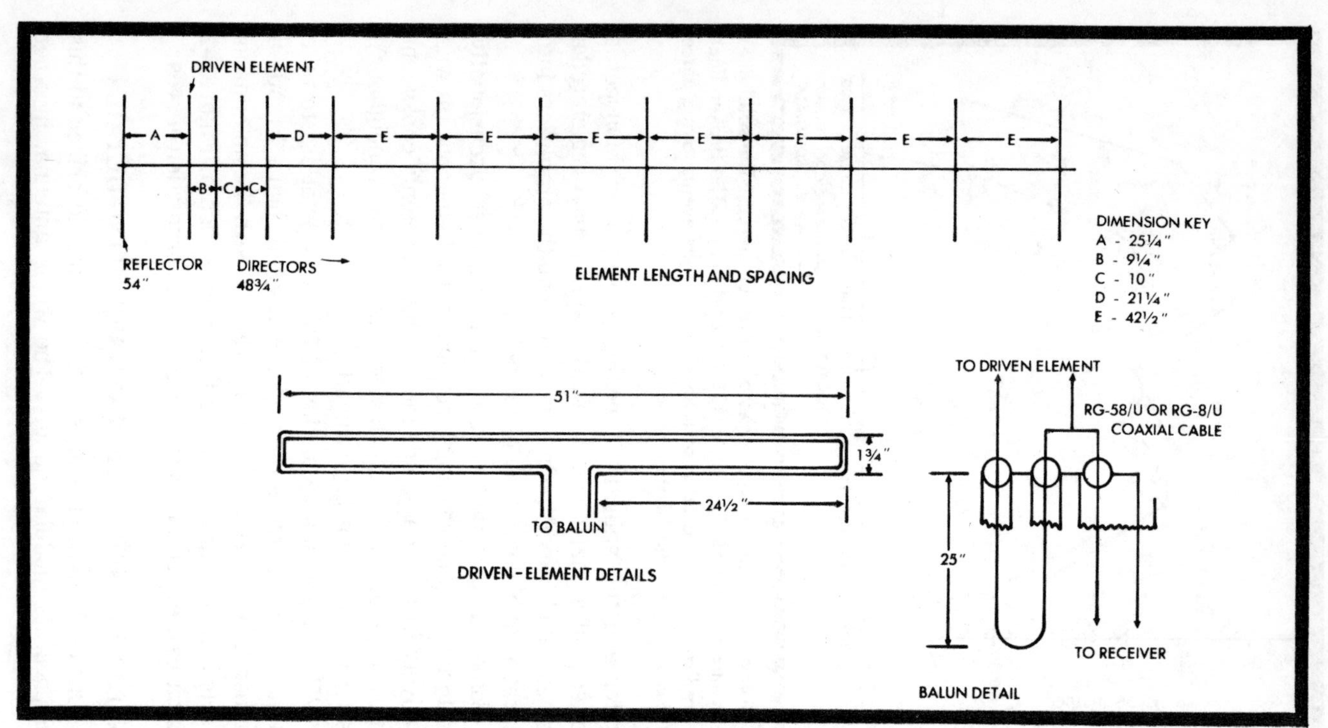

Fig. 8-3. Electrical details of the 13-element, 110 MHz beam antenna.

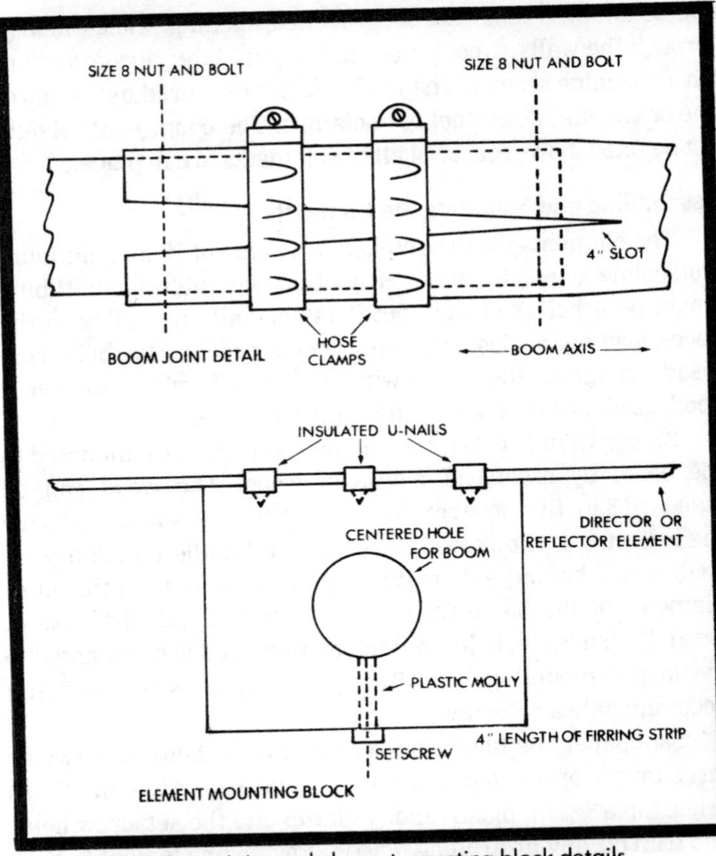

Fig. 8-4. Boom joint and element mounting block details.

words, is designed so that it can be readily disassembled and reworked to operate in a different frequency range.

Assembling the Boom

The boom can be made up of several lengths of electrical metallic tubing, top rails for chainlink fence, or any other kind of rigid, light-but-durable tubing that has an outside diameter of 1–1½ in. The total length of the boom must be at least 31 ft, but the number of lengths of tubing should be kept as small as possible.

As shown in Fig. 8-4, the lengths of tubing are joined by sawing a 4 in. slot up one end of each tube, then inserting one end of another tube into the widened opening. Holes drilled through the walls of both tubes at the joint later accommodate No. 8 machine screws. The joint is further secured by means of one or two stainless steel hose clamps. The boom joints should not be fixed, however, until all the elements are in place.

Assembling and Mounting the Elements

The elements for the antenna are made of ⅛ in. aluminum clothesline wire. Regular ⅜ in. thin-wall aluminum tubing would be a better choice, but it is normally available in the necessary quantities only on special order. The only real disadvantage of the ⅛ in. wire is that it bends whenever a good-sized bird perches on one of the elements.

As shown in Fig. 8-4, the antenna elements are mounted to the boom by means of blocks of wood. The wood can be standard 3 in. firring strip stock that is cut into 4 in. lengths for the reflector and directors, and to 6 in. for the folded dipole. Drill a hole having a diameter slightly greater than the outer diameter of the boom through the center of all 13 blocks of wood. Drill a 3/16 in. hole from the bottom of each block and into the larger mounting hole. The 3/16 in. holes will eventually accommodate a setscrew.

Sandpaper the blocks of wood and coat them with two or three layers of outdoor paint or spar varnish. After the finish dries, tap a 3/16 in. plastic molly clamp into the setscrew holes and trim off any protruding plastic. Fasten an element to each of the blocks with several insulated hammer-in staples.

Fit together the boom sections temporarily, and score the beam where the elements are to be placed. (See Fig. 8-3.) Disassemble the boom and slide the appropriate elements onto the tubing from the uncut ends. Although the elements do not have to be fastened to the boom at this point, it is important to have the proper type and number of elements on each length of boom before fastening it together. The element mounting blocks cannot be slid across a boom joint.

With the elements thus placed onto the boom sections, fit together the sections, securing the joints with sheet metal screws and hose clamps. Tighten each of the element mountings just enough to keep them from sliding and rotating on the boom.

Assembling the Mount

The principal parts of the mounting yoke include a mounting tee, a pair of beam supports, a main mounting pin, and a counterweight. The tee section is made of 2×2 stock lumber that is cut and nailed together as shown in Fig. 8-5. The two beam supports are made up of electrical conduit; their purpose is to prevent the ends of the boom from drooping under its own weight. The counterweight assembly sets the antenna's center of gravity on the mast.

After cutting the 2×2 sections for the mounting tee, drill a ⅜ in. hole near the top of the vertical member for the main mounting pin (Fig. 8-5). Sandpaper the tee sections and weatherproof them with outdoor enamel or spar varnish. When the finish is dry, nail together the sections with aluminum nails.

The beam supports can be ½ in. electrical conduit that is between 8 and 10 ft long. Drill a $^{3}/_{16}$ in. hole about 2 in. from one end of both beam supports, and mount them to the tee section. The two beam supports should pivot freely in the mounting tee so that the point of support on the boom can be adjusted.

The main mounting pin is a length of ⅜ in. threaded steel or hardened aluminum bar stock. The length is not critical so long as it is more than 12 in. The counterweight can be any 10–15 lb object that can be threaded onto the main mounting pin. I made a suitable counterweight by filling a flower pot with concrete.

To make a flower pot counterweight, find a medium-sized tile flower pot that has a hole about ½ in. across in its bottom. Cut a piece of 2×2 lumber to a length just equal to the depth of the pot. Center the block upright in the pot and, while molding the wood securely in place, fill the pot with a mixture of concrete and stone. When the mixture hardens, use the hole in the bottom of the pot as a guide and drill a ⅜ in. hole all the way through the block.

To make the beam easily adjustable in the altitude plane, it must be fastened to the boom mounting yoke with its center

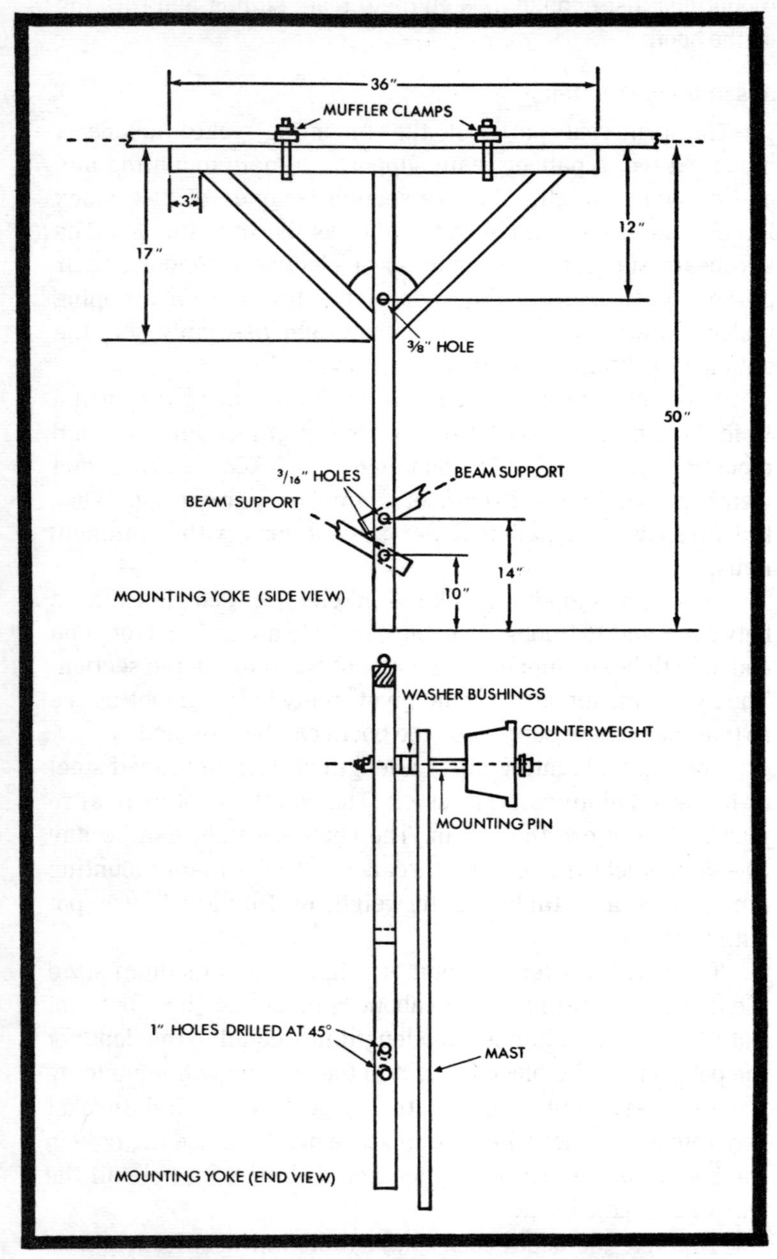

Fig. 8-5. Mounting yoke and mast attachment details.

of gravity directly over the vertical member of the support tee. Locate the center of gravity of the boom by lifting it near its geometric center and sliding it back and forth in the hand until it balances easily. Score this balance point on the boom tubing.

Find the center of the horizontal member of the mounting tee, and rest the beam on it. Adjust the boom so that the center-of-gravity mark falls directly over the center of the horizontal member on the tee. Secure the beam to the yoke with at least two 2 in. U-bolts or muffler clamps.

Slide the main mounting pin through the ⅜ in. hole in the vertical member of the mounting tee, and secure it as shown in Fig. 8-5. The two boom supports can be fastened to the boom at this point in the operations; but it is better to postpone final adjustments until the whole assembly is mounted to the mast.

Erecting the Antenna

Whereas smaller antennas can be mounted to a chimney or strapped to the side of a one-story building, the long beam antenna described here must be free-standing to permit full freedom of motion in the altitude plane. Perhaps a salvaged utility pole would make the best mast for this antenna structure—I used standard 2 in. steel mast stock for the sake of convenience.

The mast can be driven about 3 ft into the earth or dropped into a hole filled with about 2½ ft of concrete. I found it more convenient to strap the bottom section of the mast to a corner post of an existing chainlink fence.

Building the mast out of several shorter sections of mast tubing has an advantage over using two or three longer sections. For one thing, it allows the antenna to be set up only 5–6 ft from the ground for making final mechanical adjustments and stability tests. When the adjustments are complete, the mast can then be reconstructed to set the beam to its full height.

With the beam mounted at head height from the ground, make final adjustments of the element spacing and horizontal alignment, and fasten all elements to the boom with long setscrews. Fix the supports to the boom to eliminate the sag at the ends, and fix the counterweight into place.

Before erecting the antenna to its final height, it is a good idea to attach the balun and transmission line to the dipole section. This added weight should make the beam a bit heavier at the dipole end, and thus make the lower end of the boom drop gently to the ground. This in a convenient off-balance condition that simplifies altitude setting during normal operating sessions.

Even though my own mast was only 16 ft high and was made of good 2 in. steel mast stock, the weight of the boom was enough to bend the upper sections of the mast under normal operating conditions. The whole assembly appeared dangerously unstable in moderate winds. The mast must be carefully guyed down, then.

The guy wires can be attached to the mast just below the bottom of the mounting tee before erecting the beam to its final height. Conventional guying techniques are suitable, providing the wires are arranged so that they do not interfere with the altitude motion of the beam.

NORMAL OPERATING PROCEDURES

My intention in designing this 110 MHz radio telescope was to demonstrate the feasibility of constructing a workable system using simple parts and equipment. In many respects, it is a crude system that offers a number of opportunities for improvement. But what is more important is the fact that the system, crude as it is, works well.

The first indication that the system was going to work came about when I set the beam to an arbitrary point in the sky and attached the transmission line to the receiver. There was no meter connected to the receiver's audio output at the time, but a distinct increase in the level of static was a sure sign the system was functioning.

The first real system tests were begun about two hours before the Sun was due to cross the meridian. I adjusted the altitude of the beam by means of a rope attached to the bottom of the tee mounting. The setting was merely an ''eyeball'' adjustment—a rough alignment of the antenna to the path of the Sun.

I adjusted the meter for the 700 mV "noise zero" reading and began taking readings at half-hour intervals. The readings hovered close to the reference level for the first 90 minutes, then began showing a steady increase for another hour. The readings leveled off for about a half hour, then dropped off gradually over another one-hour interval.

The Sun had definitely registered its VHF signals on the system, and the fact that it took an hour and a half to cross the beam was a good indication that the horizontal beamwidth was close to the predicted 30°.

I took readings at half-hour intervals for six hours the next morning. The results are shown in Fig. 8-2, and the system was credited with "DXing" the Sun, Jupiter, and Sagittarius A. I confirmed the results by observing the same parade of objects four days in a row.

Since the system demanded constant attention as far as gathering data is concerned, I put off further experiments for the time being. I did, however, adopt the habit of letting the system run on a 24-hour basis, checking the antenna reading at irregular intervals. The readings displayed the normal background noise most of the time, and the only responses of any interest were occasional voice transmissions of unknown origin. On one occasion, the meter showed a distinct increase in output level, and the sounds from the speaker were definitely not indicative of man-made radio noise. Checking a celestial map, I found the system was responding to one of the weaker radio sources—*Virgo A*.

Cassiopeia A and Cygnus A have since been added to the logbook for this simple 110 MHz system, and I've used it to make a partial mapping of the Milky Way.

It turns out that this system has a secondary application that might be of interest to avid broadcast-band DX fans—it makes a beautiful system for listening to faraway commercial FM stations.

ADDING A PREAMPLIFIER TO THE SYSTEM

Inserting an FM (88–108 MHz) preamplifier between the transmission line and receiver boosts the overall sensitivity of the radio telescope. Preamps do add some noise to the system, however, and an experimenter must be prepared to deal with the added noise by readjusting his sensitivity estimates or by exchanging the preamp for a better one.

Commercial FM Preamplifiers

FM preamplifiers combining FM and television preamps are commonly found in electronics supply stores. The prices range from about $10 to $100, depending upon power gain, noise level, and number of outputs (signal splitters). Price, however, is not an altogether reliable indication of quality, and the experimenter would be wise to buy a unit from a store that allows exchanges.

The performance of a commercial FM preamplifier also depends heavily upon how the unit is connected into the system. The beam antenna described in this chapter, for instance, presents a rather unusual 50Ω impedance to the receiver—most FM preamplifiers have connections for 75Ω or 300Ω. One of the most important considerations in connecting a commercial preamp to the system described in this chapter is thus insuring good impedance matching all along the line.

Suppose a commercial preamp has a 300Ω input and output. That is a common configuration for low-cost indoor versions. In such a case, the experimenter has to insert a 50-to-300Ω matching transformer between the transmission line and preamp input. Since ready-made matching transformers of this value may be difficult to find, the user will have to resort to an antenna tuning assembly such as Radio Shack's 21-510 variable antenna matcher.

Better preamps have 75Ω inputs and outputs. In-line 75-to-300Ω matching transformers (baluns) are readily available for about $2. Such transformers have the appropriate kinds of line connectors. A matching balun that transforms the 50Ω impedance of the antenna line to 75Ω for the preamp input are a bit more difficult to find nowadays, but they are usually available through ham specialty shops. The antenna tuner described in the previous paragraph could serve this particular function as well.

A Low-Noise 110 MHz Wideband Preamp

The homemade 110 MHz preamplifier shown in Fig. 8-6 accepts a 50Ω line impedance at its input and delivers an amplified signal at its 300Ω output. The circuit uses FET amplifiers to keep the noise level as low as possible. The gain of the preamp is about 20 dB, but the actual gain can vary either way according to how well it is constructed and aligned.

Signals from a 50Ω coax connector are fed to L1, a tapped coil that serves as both an impedance matching transformer and the inductive half of a tuned circuit. The input is coupled to the first amplifier, Q1, through C2.

The signal amplified by Q1 appears across the LC circuit composed of L3 and C5. Capacitor C4 makes the circuit appear as a common-source amplifier to rf signals. The neutralization circuit, composed of C3 and L2, is adjusted to achieve optimum balance between circuit stability and gain.

The signal is coupled from L3 to L4 by means of transformer action. The two tuned circuits are adjusted for slightly different resonant frequencies in order to achieve the wideband feature of the amplifier.

The amplifier made up of Q2 and its associated components is almost identical to the first amplifier stage. Its output is developed across a resonant load made up of C11 and L6, and finally sent to the output through the adjustable capacitor C12. C12 is fine-tuned to make the best possible impedance match between the preamp circuit and receiver input.

The entire circuit can be laid out on a small PC board. It is absolutely essential that the copper barriers be installed between the amplifier stages. The finished unit can be installed in a standard aluminum enclosure.

The power supply for this preamp can be a combination of batteries adding up to 12V.

The circuit should be aligned with a sweep generator to get the best balance between gain and bandwidth. In any event, the bandwidth must be at least 2 MHz between half-power points. Adjusting the neutralization coils, L2 and L5, is a matter of trading off gain for the circuit's stability and

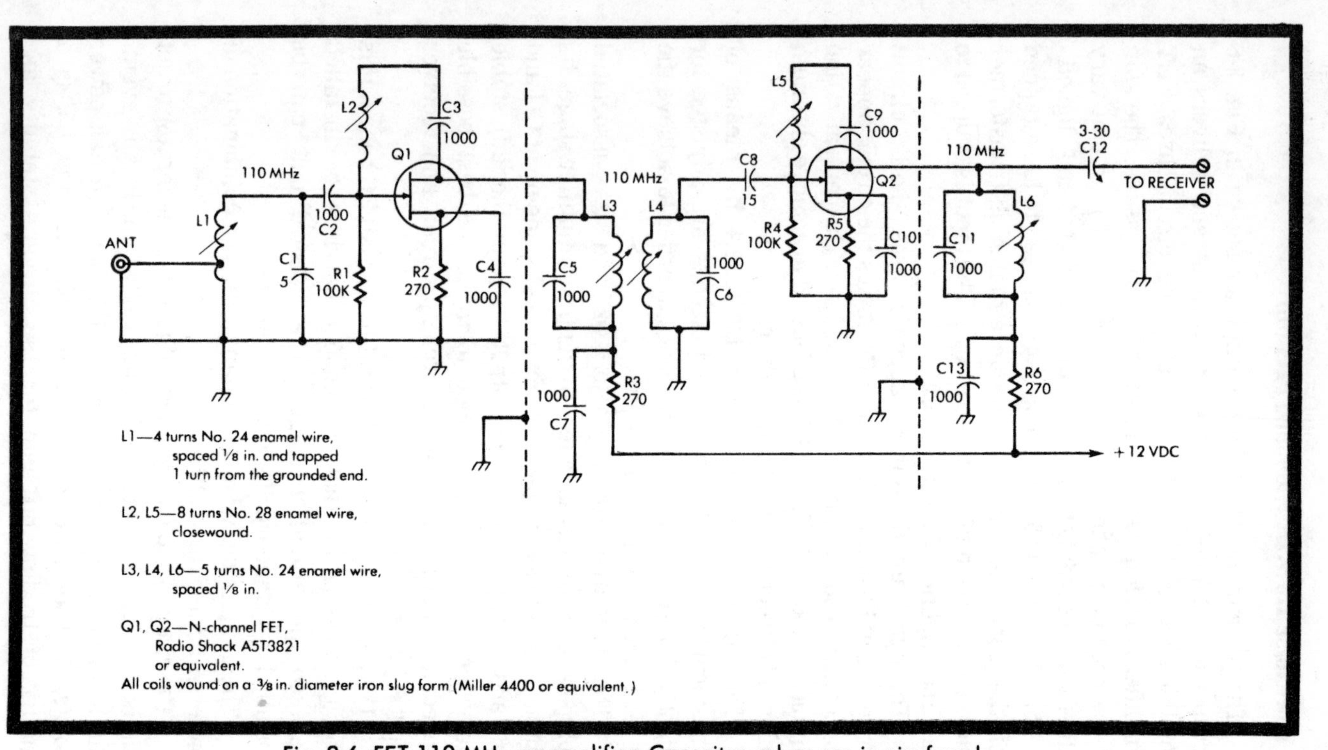

Fig. 8-6. FET 110 MHz preamplifier. Capacitor values are in picofarads and resistor values are in ohms.

inherent noise level. Adjusting the position of the tap on the input coil (L1) also has some influence upon the circuit noise level.

The circuit can be tested and aligned reasonably well by connecting it between the transmission line and receiver, and tuning the receiver for a fairly weak station at the upper end of the FM band. If a sweep or signal generator is not available for initial alignment, adjust the coils for the strongest, best quality reception.

WIDEBAND 100 MHz ANTENNA FARM

The newness of amateur radio astronomy often makes it difficult for a novitiate to decide exactly what projects he wants to perform. The information and suggestions in Chapter 6 can help an amateur set up a station for "get acquainted" experiments; but serious amateurs must think about possible future projects before setting up a permanent station. Unless an experimenter is more interested in assembling radio telescopes than doing radio astronomy research, he would be wise to make his permanent station as versatile as possible.

Most of the effort involved in setting up a radio telescope goes into building the antenna system. The more versatile the antenna system, the easier it is to adapt the telescope to different kinds of projects in the future. The antenna array described in this chapter was designed to meet this need for system flexibility. The versatile nature of the system arises from its modular construction—it is made up of combinations of identical or near-identical components.

The basic antenna system component suggested in this chapter is an 8-turn helical antenna cut to a center frequency of about 100 MHz. The most useful array is made up of four such antennas, two being wound clockwise and two wound counterclockwise.

The fact that the antennas are wideband helicals means the experimenter can adjust the system's operating frequency to avoid man-made radio transmissions without sacrificing any great amount of antenna gain. By the same token, he can take advantage of the wideband nature of the antennas to check the intensity of sources at several different wavelengths.

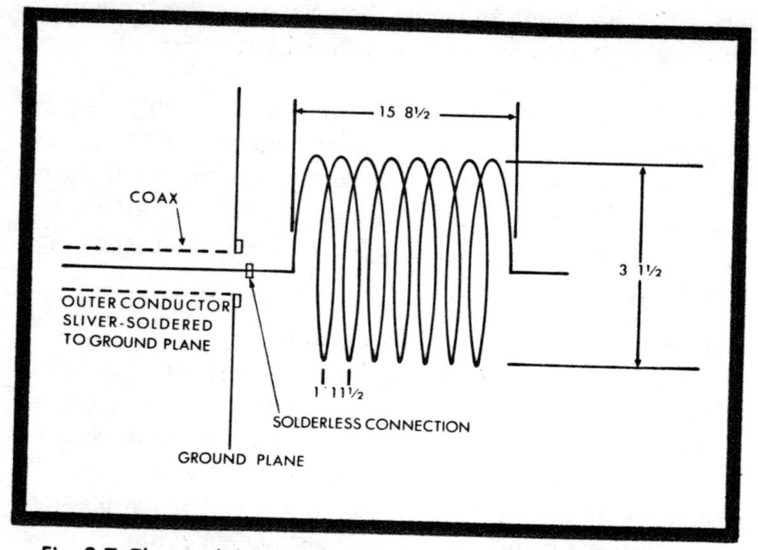

Fig. 8-7. Electrical details for a high-gain 110 MHz helical antenna.

Any number of these antennas can be ganged together to produce an antenna farm having as much gain as desired. And by winding the helicals in different directions, the experimenter can carry out some fairly sophisticated projects dealing with signal polarization. Of course, two identical sets of antennas separated by 20 wavelengths or more give the experimenter a workable interferometer system.

The Basic Antenna Element

Figure 8-7 shows the dimensions of a clockwise helical element, and Fig. 8-8 shows the construction details for the backplane and mounting frame. To keep the antenna cost as low as possible, I wound the helical elements from ordinary ⅛ in. aluminum clothesline wire and built the frame from 1×3 in. firring strip stock. The backplane reflector is made of common aluminum window screen. The beamwidth of a single helical element is on the order of 40° (half-power points of the central lobe) and the gain is about 13 dB. The characteristic impedance is about 140Ω.

Cut the firring strips for the framework and helix support member to the dimensions shown in Fig. 8-8. Firring strip

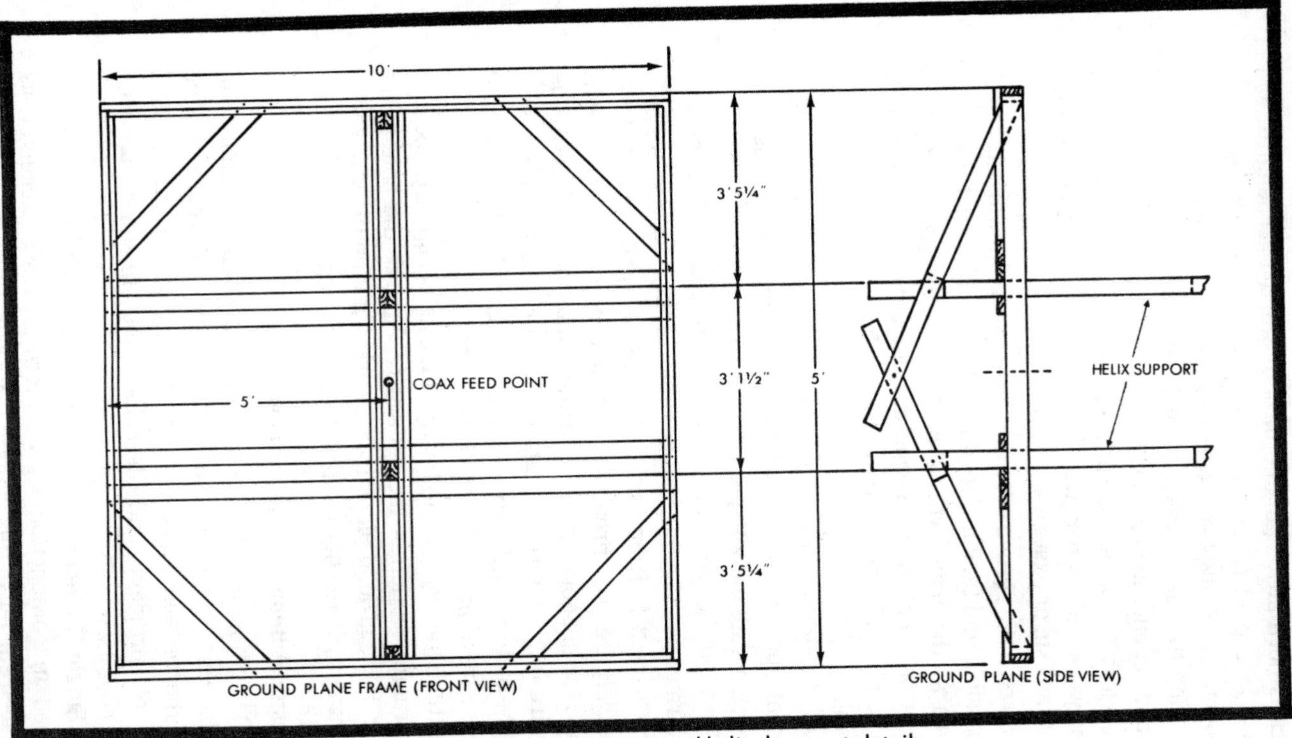

Fig. 8-8. Ground plane frame and helical support details.

stock is normally available in both 8 and 12 ft lengths. It is better to use 12 ft lengths wherever necessary to avoid the problems of joining two pieces together.

Align the long sections of helix support as shown in Fig. 8-8, and cut out the ¼ in. notches.

Assemble the framework and support sections. Coat the framework and support with several layers of outdoor enamel, spar varnish, or paste wax.

Stretch the screen over the top of the framework as tightly as possible and tack it into place. Cut a ½ in. hole through the screen at the point where the helix is to pass through the framework, and smear a generous amount of varnish over the hole to prevent frayed screen from shorting out the antenna assembly.

A simple altitude mounting scheme for the antenna element appears in Fig. 9-4. The experimenter can devise a more elaborate mounting scheme if necessary.

To wind the helix, cut an 82 ft section of clothesline wire and straighten it as much as possible. Make a right-angle bend in the wire 2 ft from one end. That particular section will eventually be fed through the ground plane. Wind the wire around the support member, fitting it into the notches. Adjust the wire for "true round" after each turn, and tack the wire into the notches with small insulated U-nails. Trim off any excess wire after making the final turn.

Attach the guy lines to the tip of the helix support, and feed the straight section of the helix wire through the hole in the center of the ground plane framework. Tighten the guy wires, making certain the helix support stands straight.

Suggested Antenna Configurations

The number of helical sections with a given combination of polarizations depends upon the kind of project the experimenter wants to carry out. This section outlines several kinds of projects and describes the modular antenna requirements.

A Single-Element Radiometer. A single helical section can be used in conjunction with an ordinary FM receiver as described earlier in this chapter. The sensitivity of the system will not be as great as the system in Fig. 8-1, because the lone helix has a much lower gain than a long beam antenna.

This helical system is nevertheless useful for detecting polarized radio emissions from the Sun and Jupiter. Jupiter's signals are more interesting in the decameter range, but this particular system gives the experimenter an opportunity to study Jupiter with a polarized antenna.

The system's overall sensitivity can be improved by connecting an FM preamplifier in front of the receiver section.

The only technical problem associated with this setup is matching the 140Ω characteristic impedance of the helix to the 300 or 75Ω input on the receiver. Since matching transformers with a 140Ω input are not normally available from retail supply houses, the experimenter must either wind his own matching transformer or use a commercial antenna tuner. In any event, it is important to maintain the unbalanced characteristic of the helix right up to the receiver input.

A Two-Section Radiometer. Constructing two basic helical sections, one wound for clockwise and the other for counterclockwise polarization, opens up several interesting avenues for amateur research.

Figure 8-9 illustrates a modified version of the single-element radiometer. In this instance, the experimenter uses an antenna switch to change the polarization of the system at will. The overall sensitivity of the system is rather low when used in this fashion, and the experimenter still has to cope with the tedious impedance matching problems.

The primary objective of such a scheme would be to study the polarization characteristics of powerful radio sources by comparing the receiver output while changing from one antenna to the other.

Using a pair of antennas switched as shown in Fig. 8-9 also lets the experimenter change the area of observation. One antenna, for example, could be pointed toward the ecliptic to observe a meridian crossing of Jupiter. The other antenna could be aimed at another powerful source such as Cygnus A. A simple turn of the antenna switch lets the experimenter effectively carry out two independent sets of observations at the same time.

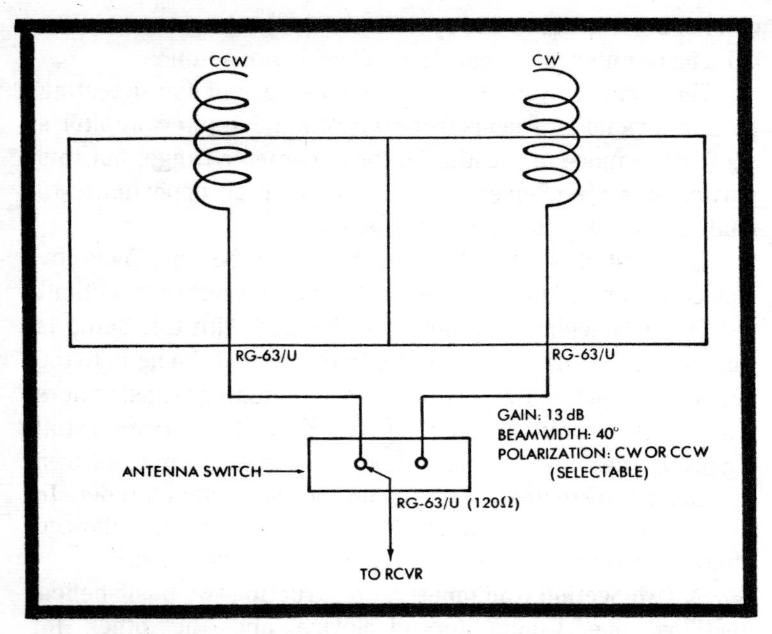

Fig. 8-9. A two-helix array for polarization studies.

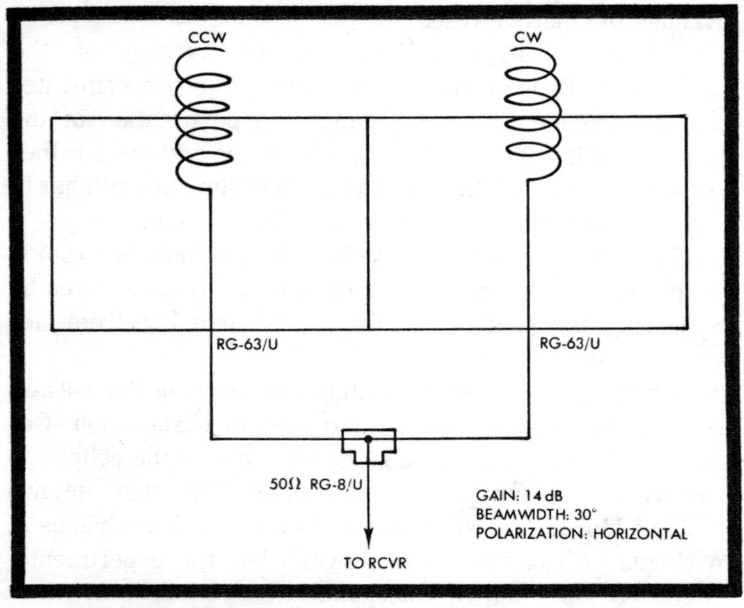

Fig. 8-10. A horizontally polarized 110 MHz helical bay.

The antennas should be located so that the presence of the unused antenna does not interfere with the lobe pattern on the one connected to the receiver.

Figure 8-10 shows another configuration for the two-antenna scheme. In this instance, the objectives are to make the favored direction of polarization linear and, at the same time, increase the overall antenna gain.

The two antenna sections are mounted side by side. It is a good idea to fasten the two frames together to insure the same altitude and azimuth angles. The antenna leads are joined with a T-fitting, producing a transmission line impedance of about 70Ω.

The fact that the two antenna sections are wound in different directions nullifies their circular polarization and changes it to a linear polarization. Unless one antenna is stacked atop the other, the polarization will be horizontal.

Interferometer Systems. The techniques employed for setting up an antenna system for an interferometer-type radio

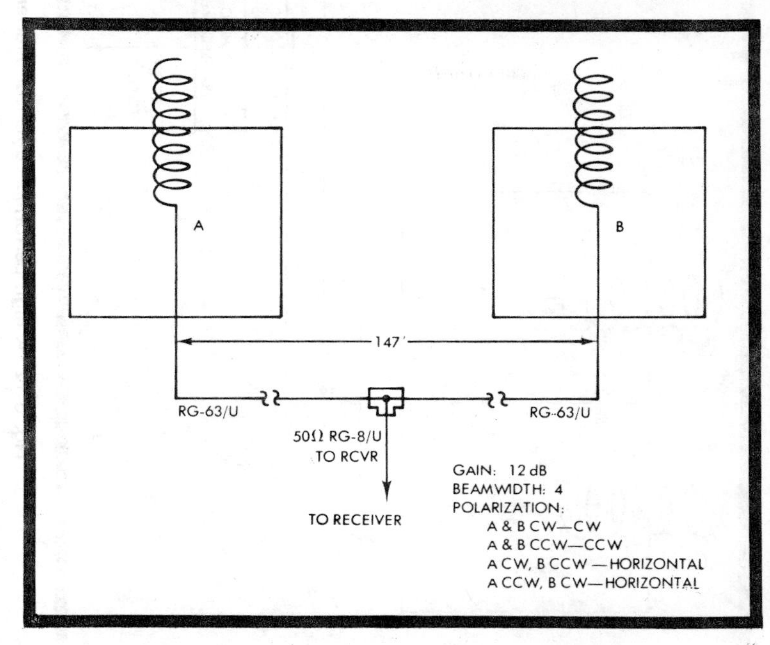

Fig. 8-11. A helical interferometer with polarization options.

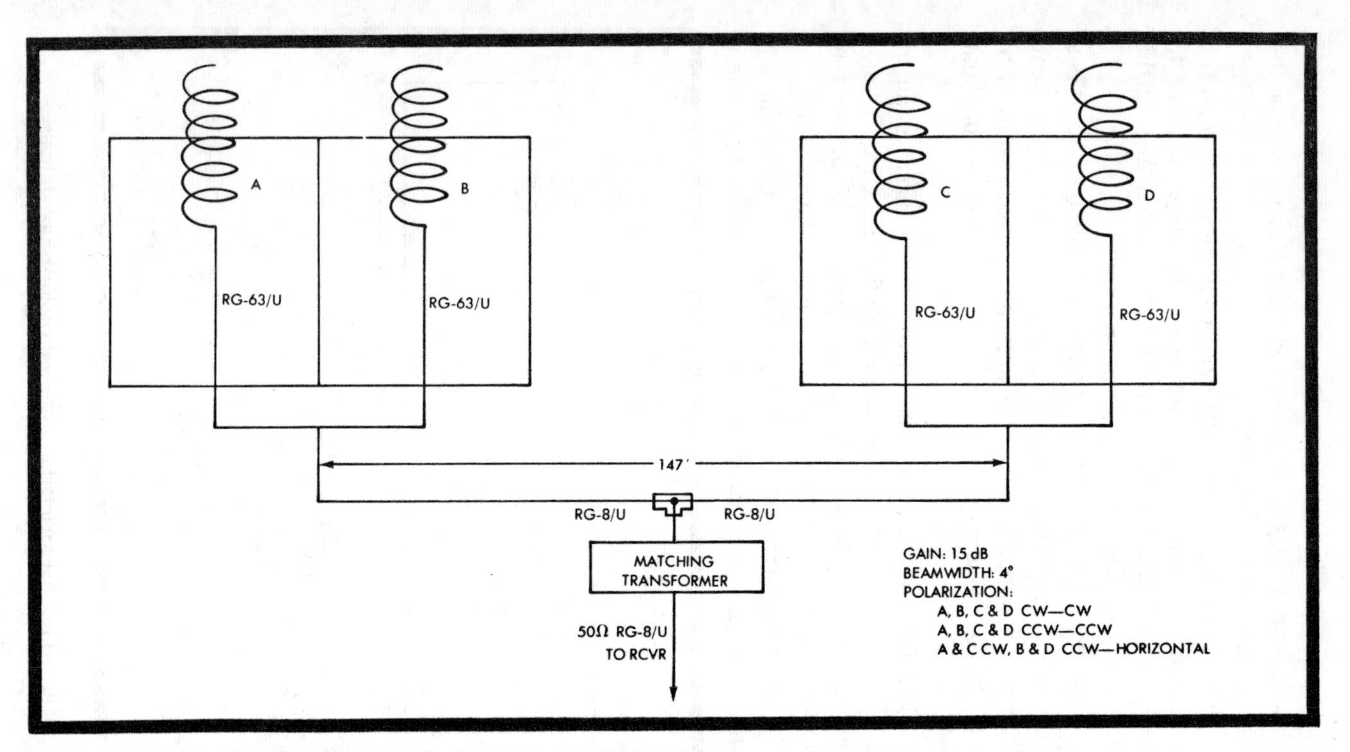

Fig. 8-12. A high-performance helical interferometer with polarization options.

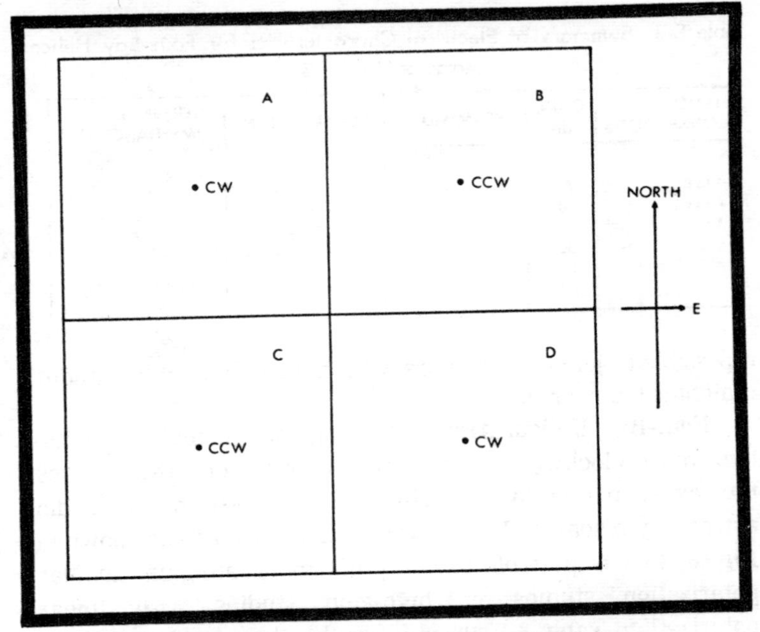

Fig. 8-13. A high-performance radiometer helical array.

telescope can be applied with the helical elements described here. The spacing between the antennas must be some even multiple of the operating wavelength, and preferably more than 15 wavelengths.

Using a pair of helixes with the same sense (both clockwise or counterclockwise) makes the system favor signals having one particular direction of circular polarization. Using oppositely turned helixes makes the system more responsive to horizontally polarized signals. (See Fig. 8-12.)

It is possible to increase the gain of the antenna system by using a pair of helical elements at both antenna stations. The antennas can be interconnected as shown in Fig. 8-13 to produce the desired system polarization.

The 4-element interferometer system, incidentally, does not present the same impedance-matching problems that plague the other schemes suggested thus far. The combined outputs of each pair of antennas is almost exactly 70Ω, making

Table 8-1. Summary of Electrical Characteristics for Four-Bay Helical Array of Fig. 8-13

ELEMENT CONNECTIONS	GAIN, dB	BEAMWIDTH	POLARIZATION	TERMINAL IMPEDANCE, Ω
A to B	30	14	Horizontal	52
A to C	30	14	Vertical	52
A to D	30	14	Clockwise	52
B to C	30	14	Counterclockwise	52
A, B, C, and D	21	17	Linear	26

it possible to join the two pairs with 72Ω coaxial cable without matching transformers.

Four-Bay Helical Array. Building four helical elements, two wound clockwise and two wound counterclockwise, gives the experimenter a powerful antenna system for radio astronomy research. If the antennas are arranged as shown in Fig. 8-14, it is possible to carry out moderate-gain circular polarization studies or high-gain studies with linear polarization. Table 8-1 summarizes the characteristics of the antenna systems based upon the arrangement in Fig. 8-14.

It is important to bear in mind that the combined outputs of any two antenna elements will have an impedance on the order of 70Ω, while the combined outputs of all four sections drops the characteristic impedance of the system to about 36Ω.

146 MHz Interferometer Systems

The closing sections of Chapter 7 describe the special advantages and general characteristics of interferometer-type radio telescope systems. The special advantage of an interferometer system is its high degree of resolution compared to a radiometer counterpart. The main constructional characteristic of an interferometer is the fact that it uses a pair of identical antennas.

Figure 9-1 shows a block diagram of the 146 MHz interferometer system. The two helical antennas are identical in all respects and they are fixed to simple altitude mountings. (See Fig. 9-4.) The gain of each antenna is about 13 dB, neglecting line losses, and the net gain of the antenna system is on the order of 15 dB. The beamwidth of the central lobe of each antenna is about 45°, but the beamwidth of the interference pattern takes the system resolution down to about 3°.

An experimenter can use either of the two basic receiver schemes suggested in this chapter. Both receivers employ a homemade front end—rf amplifier, mixer, and local oscillator. The intermediate frequencies for the two receivers are different, however: One uses a homemade or commercial i-f strip tuned to 30 MHz, and the other uses the MHz i-f section from a conventional TV set.

An integrator and dc amplifier couple the output of the i-f section to the recorder. The time constant of the integrator is

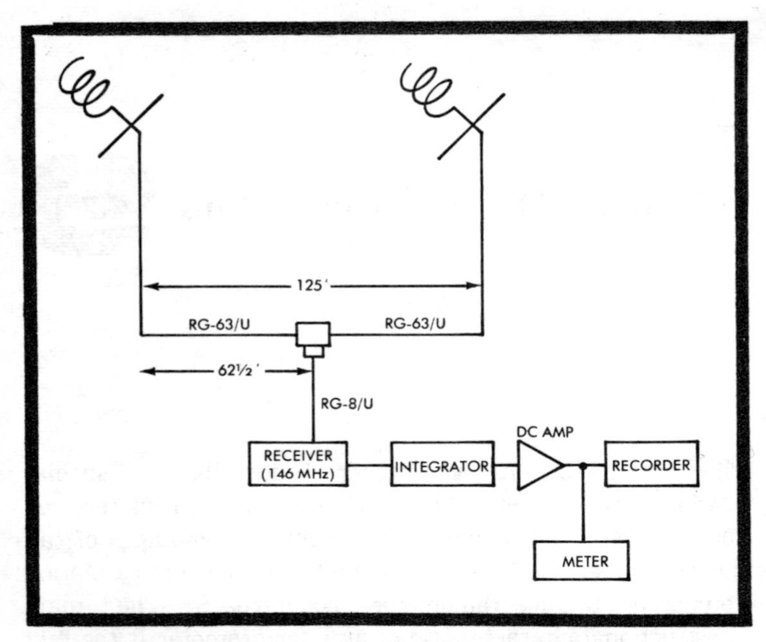

Fig. 9-1. Block diagram of the 146 MHz interferometer system.

adjustable, and the dc amplifier gain is variable between 0 and about 48 dB.

The readout device can be either some sort of voltmeter or a chart recorder. The discussions in this chapter assume the user has access to a chart recorder, however.

The system as it is described here is capable of reliably detecting radio signals from the Sun, Jupiter, Cassiopeia A, and Cygnus A. Taurus A, Virgo A, and Sagittarius A can be added to the list of detectable sources after making suitable modifications in the basic antenna system and the data-gathering procedures.

THE BASIC ANTENNA SYSTEM

Figure 9-2 shows the dimensions for the basic antenna element. It is a 7-turn helix that is cut for a center frequency of 146 MHz and wound in a clockwise direction. The direction of the winding is not critical in this instance, but the dimensions are.

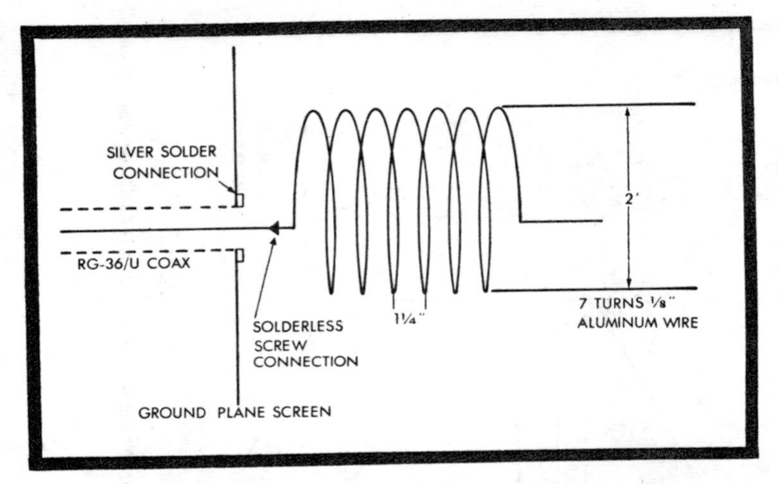

Fig. 9-2. Electrical details of 146 MHz helix—two identical antennas are required.

The frame is made of 2×2 lumber and 1×3 firring strip stock. The wood must be coated with several layers of spar varnish, outdoor enamel, or floor wax—preferably before assembling the antenna structure. The helix support is made of furring strip stock that is weatherproofed in a similar fashion. As in the case of the 100 MHz helicals described in Chapter 8, the basic 146 MHz helixes are wound from common ⅛ in. aluminum clothesline wire. The ground plane reflector is ordinary window screen. (See Figs. 9-3 and 9-4.)

A pair of identical antennas are separated by a distance of 125 ft and always aimed at the same point in the sky. Since the helicals have a characteristic impedance of about 140Ω each, they should be connected together with RG-63/U (125Ω) coax. The coax must be cut in its exact center, or 62½ ft from the ends. A T-connector at the junction point will show about 60Ω impedance, and the combined signals can be carried to the receiver by means of ordinary 50Ω coax.

THE RECEIVER SYSTEM

Figure 9-5 compares the block diagrams of the two receiver systems suggested in this chapter. The receiver illustrated in Fig. 9-5A produces a 30 MHz i-f that can be

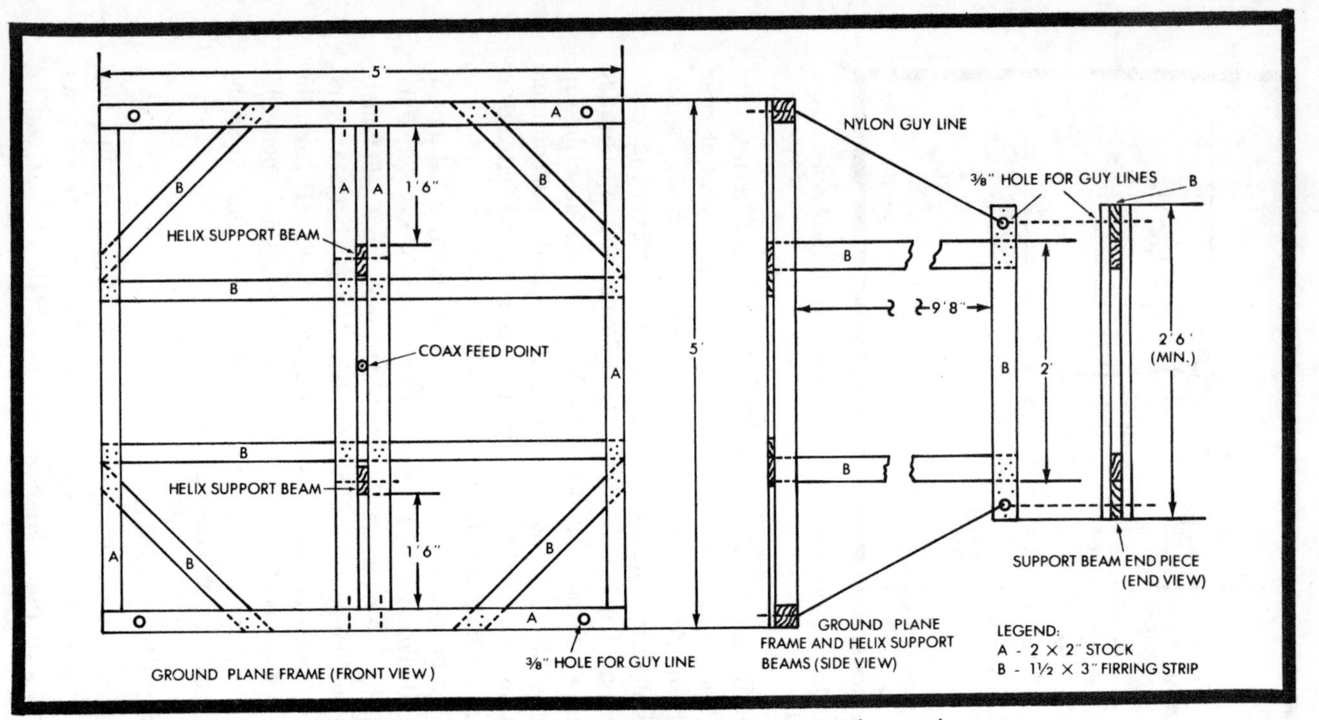

Fig. 9-3. Mechanical details of the 146 MHz helix ground plane and support.

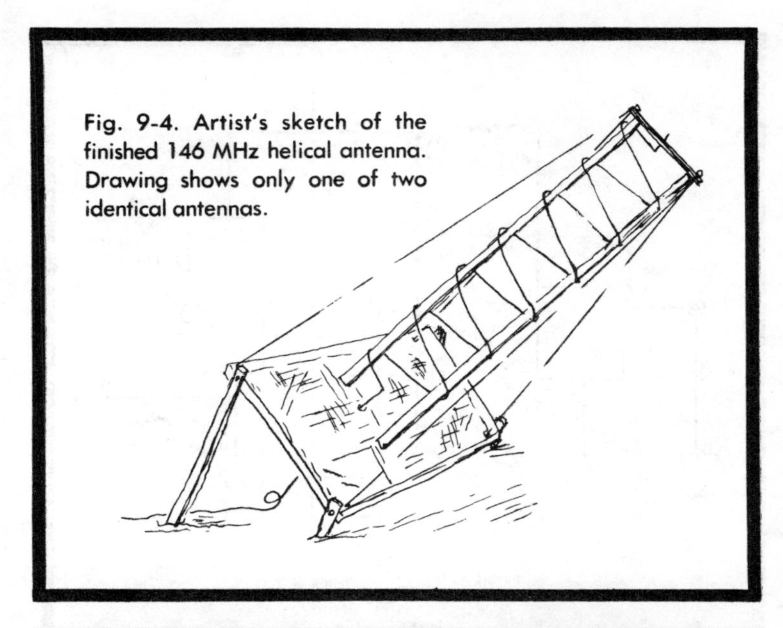

Fig. 9-4. Artist's sketch of the finished 146 MHz helical antenna. Drawing shows only one of two identical antennas.

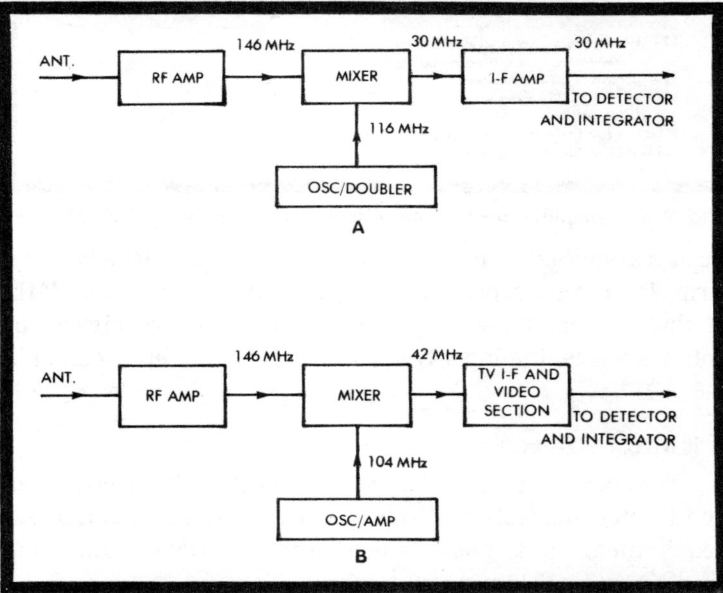

Fig. 9-5. 146 MHz receiver options: (A) 30 MHz i-f unit; (B) 42 MHz i-f unit.

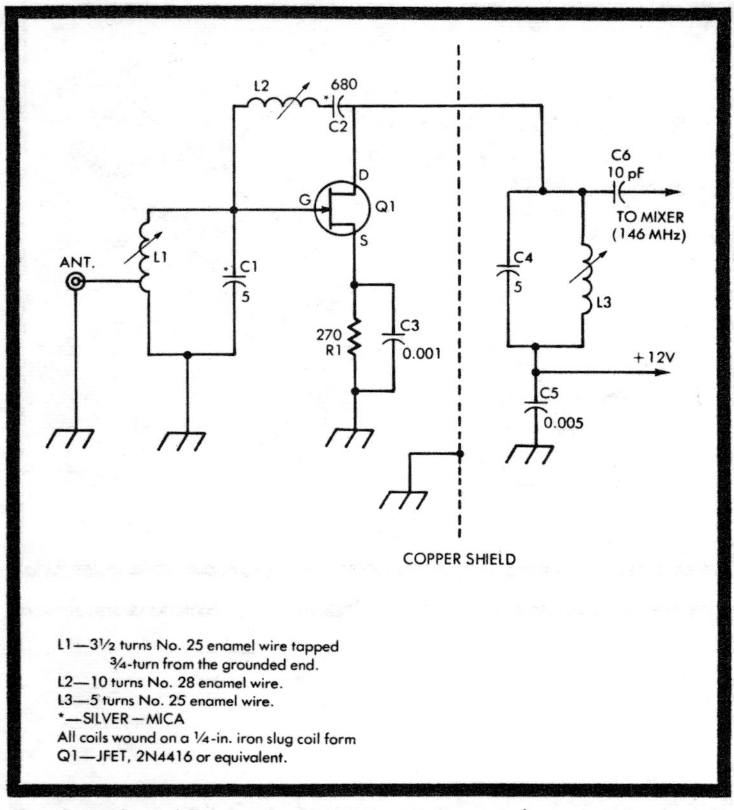

Fig. 9-6. Rf amplifier for the 146 MHz receiver employing a 30 MHz i-f.

further amplified by a homemade or commercial wideband i-f strip. The circuit represented by Fig. 9-5B produces a 42 MHz i-f that is compatible with common television receivers. In both instances, the input frequency to the front-end section is 146 MHz.

A 30 MHz I-F Receiver

The receiver system that is the subject of this section was constructed and tested by Robert R. Kenyon as part of a thesis requirement for a master's degree in electrical engineering from Ohio State University.

As shown in Figs. 9-6 through 9-8, the receiver front end is a basic amplifier—mixer combination that uses a crystal and frequency doubler to provide a local oscillator frequency of 116 MHz.

Referring to Fig. 9-6, the antenna signal is applied directly to slug-tuned coil L1, which serves the double function of an impedance matching autotransformer and the inductive element of an LC tuned circuit. L1 and C1 present a high impedance to the gate–source circuit of Q1 at 146 MHz. The rf amplifier, made up of Q1 and its associated components, is dc biased by means of a small source resistance (R1) and the source-to-drain current flow through L3. A neutralization circuit composed of L2 and C2 has to be adjusted for an optimum balance between amplifier stability and gain.

Source bypass capacitor C3 lets the circuit work like a high-gain, common-source amplifier at radio frequencies. A copper shield between the FET amplifier and the output circuit reduces rf feedback from the mixer. Bypass capacitor C5 performs the same kind of function by eliminating any rf picked up in the supply conductor.

The amplifier's output circuit, made up of a parallel tuned circuit and a signal-coupling capacitor, provides a maximum load impedance for the mixer (Q2) at 146 MHz and passes the signal to the mixer stage.

The oscillator circuit shown in Fig. 9-7 derives a 58 MHz signal from crystal Y1, and amplifies the oscillations by means of an FET circuit composed of Q3 and an rf load circuit (C13 and L7) tuned to about 58 MHz. The oscillator signal is further amplified by Q4. The parallel-tuned load for the FET (C17 and L8) is tuned to the oscillator's 116 MHz overtone.

The output of the frequency doubler is finally coupled to the mixer stage by means of transformer action between L8 and L9. Like the L8 primary side, the L9 secondary is tuned for 116 MHz resonance.

A 9.1V zener diode stabilizes the frequency output of the 58 MHz portion of the circuit by smoothing out power supply voltage fluctuations. The diode is not absolutely essential if the receiver is operated from a battery supply. The low-pass filter circuit (RFC1 and C15) prevents any 85-to-116-MHz feedback through the power supply conductors.

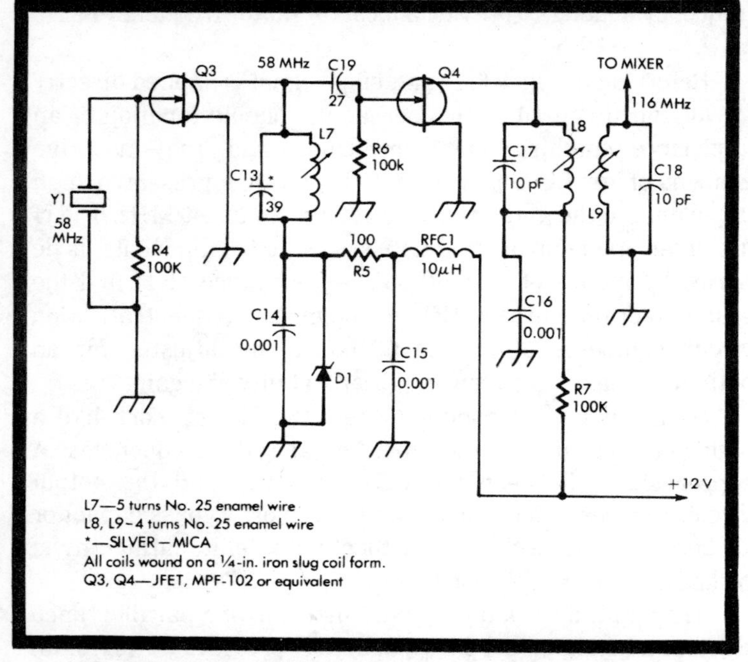

Fig. 9-7. Oscillator—doubler section for 30 MHz i-f system.

The mixer circuit, illustrated in Fig. 9-8, accepts 146 MHz signals from the rf amplifier and a steady 116 MHz oscillation from the oscillator–doubler circuit. Using one of the simplest and most popular mixer schemes available today, the circuit combines the two signals in a MOSFET to produce a difference frequency of 30 MHz.

The signal from the oscillator–doubler is coupled through C8 to gate 2 of Q2. L4 and C7 provide a high-impedance load at 146 MHz, and thus effectively couple the rf signal to gate 1 of Q2.

Transistor Q2 mixes the two incoming rf signals and amplifies the 30 MHz difference frequency. The other heterodyned frequencies, namely a 262 MHz *sum* frequency and the two original input frequencies, are not amplified by this circuit, because the tuned load (L5 and C10) appears as a low impedance to them. The 30 MHz i-f is coupled to the i-f stage by means of transformer action between L5 and L6.

A homemade version of the i-f strip shown in Fig. 9-9 is composed of three IC amplifiers that are capacitively coupled and tuned to 30 MHz with LCR feedback elements. Kenyon's final model, however, used a commercial i-f strip, Model IF-31-BS, available from LEL Products (Varian Solid-State Div., Akron St., Copiague, N.Y. 11726). A 30 MHz surplus radar i-f strip is also suitable for this application.

The commercial and surplus i-f strips have to be detuned and realigned to provide a gain of 80–90 dB at 30 MHz. The bandwidth should be on the order of 600 kHz.

The power supply for the front end and homemade i-f strip can be a standard regulated +12V supply or a pair of 6V batteries connected in series. The LEL i-f strip is of the vacuum-tube variety, and so calls for a separate B+ supply.

Construction Procedures. The experimenter can build up each of the circuits in Figs. 9-6 through 9-9 on a separate circuit board or place the entire system on a single board. Since the layout of the components can be critical as far as

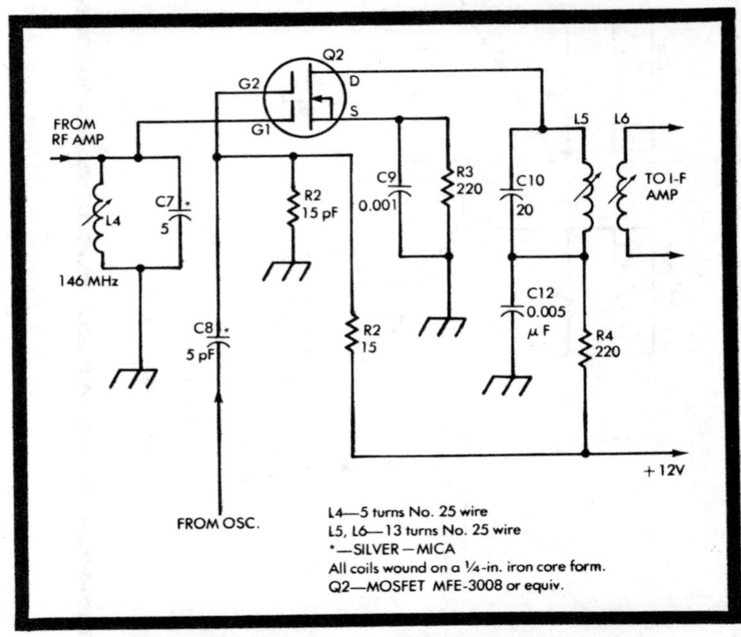

Fig. 9-8. Mixer section for the 30 MHz i-f receiver.

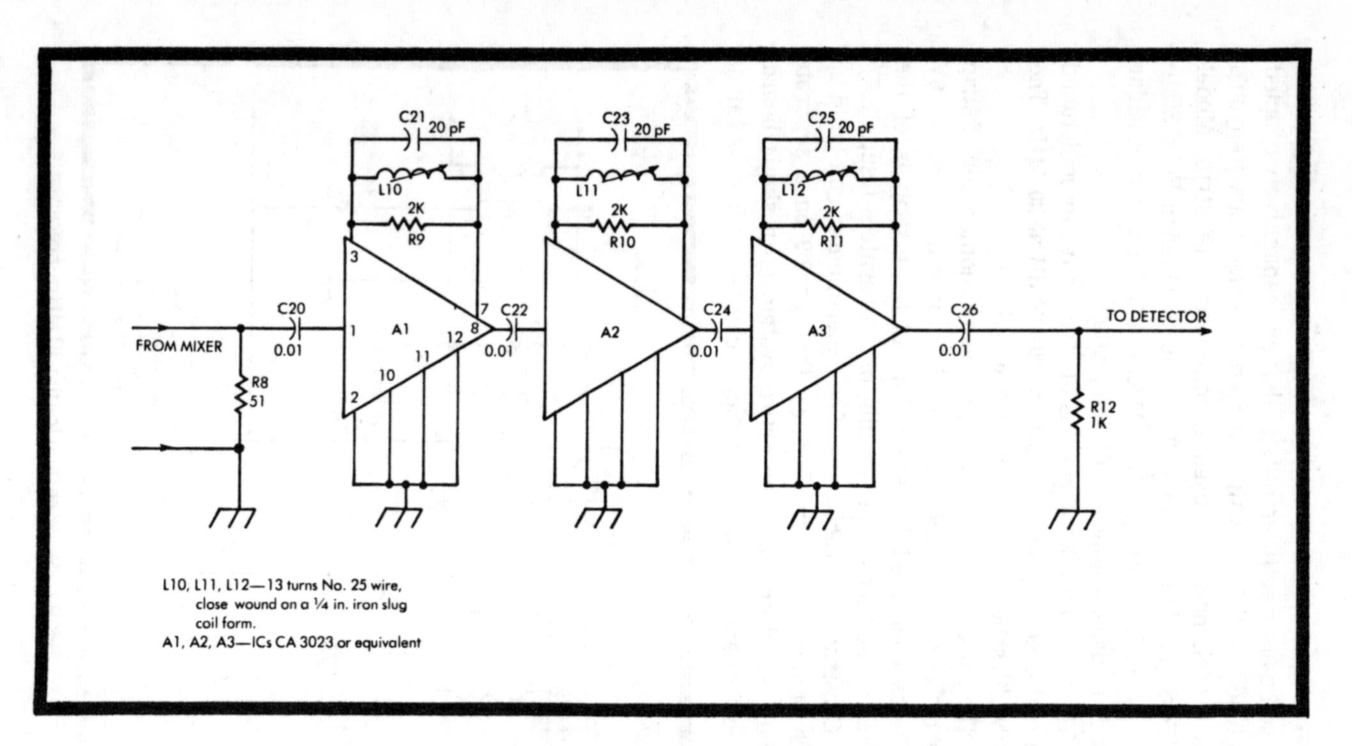

Fig. 9-9. A suggested 30 MHz i-f amplifier section for the 146 MHz radio telescope.

unwanted feedback is concerned, it is perhaps better to assemble the rf amplifier, oscillator, and mixer stages on one board, put the i-f strip on another, and assemble the integrator circuit and dc amplifier (described later in this section) as separate units. This particular procedure allows the experimenter to adjust the physical positions of the basic stages to avoid unwanted feedback.

Initial Adjustments and Alignment Procedures. The i-f section, whether homemade or purchased in finished form, should be aligned first. The homemade version suggested in Fig. 9-9 will most likely generate oscillatory signals at first, and the experimenter can expect to spend some time and effort isolating the individual stages before alignment is possible. Using a sweep–marker generator, tune the i-f transformers and chokes for a center frequency of 30 MHz and half-power bandwidths of about 600 kHz. Since the gain of the i-f strip has to be on the order of 90 dB, the input-to-output signal voltage ratio has to be about 1:1000.

Adjust L7 in the local oscillator circuit for peak 58 MHz response across the coil, and adjust the slugs in L8 and L9 for a peak oscillator output of 116 MHz across C18.

Adjust L2 in the rf amplifier to minimize unwanted oscillation in that section. Inject a 146 MHz sweep–marker signal at the antenna terminals, and adjust the tuning slugs in L3 and L4 for the best possible rf response at C7. Since the gain of the rf section has to be close to 12 dB, the input-to-output voltage ratio should be about 1:4. The bandwidth should be no less than 600 kHz between half-power points.

A 146 MHz Converter for 42 MHz I-F

The circuits illustrated in Figs. 9-10 through 9-12 serve as a front-end for a wideband 146 MHz radio telescope receiver. The i-f strip in this instance can be that of an ordinary TV set. The pickoff point for the output signal can be at the TV's i-f or video output point.

The front-end circuits—rf amplifier, oscillator–doubler, and mixer—are almost identical to the front-end section for the 30 MHz i-f version. The major differences between the two

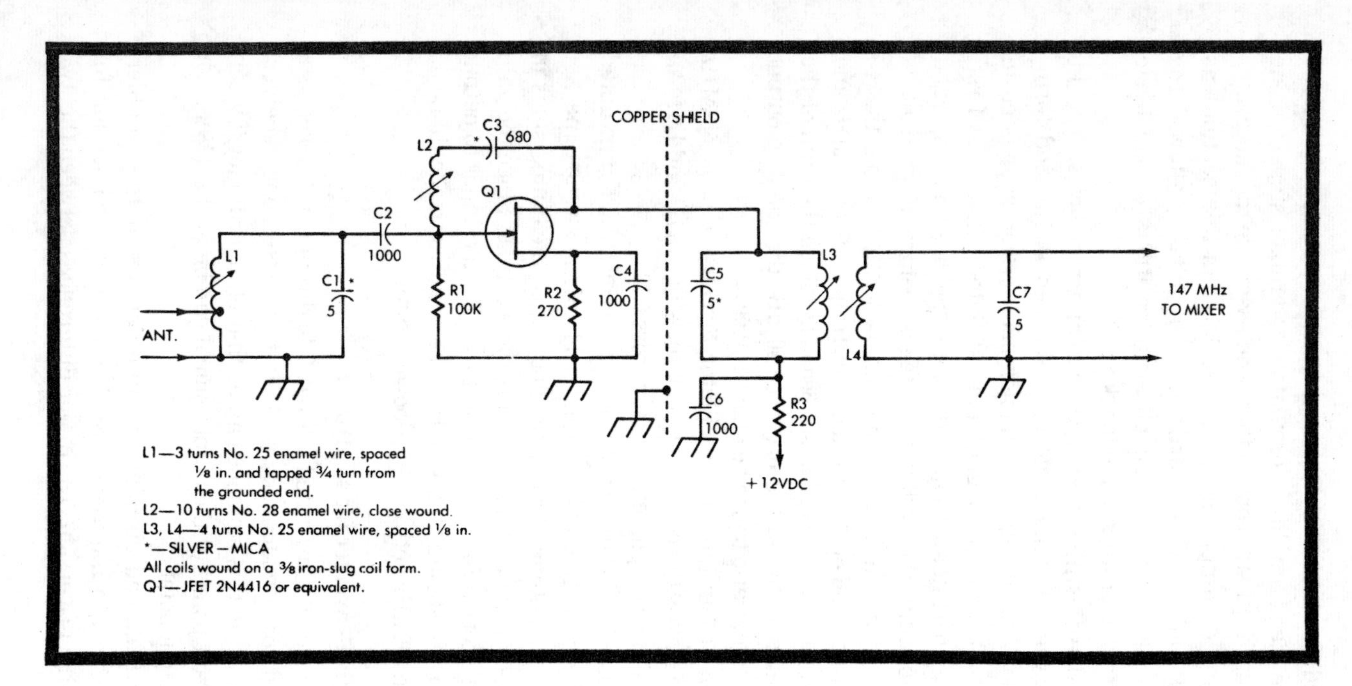

Fig. 9-10. Rf amplifier for the 146 MHz receiver employing a 42 MHz i-f.

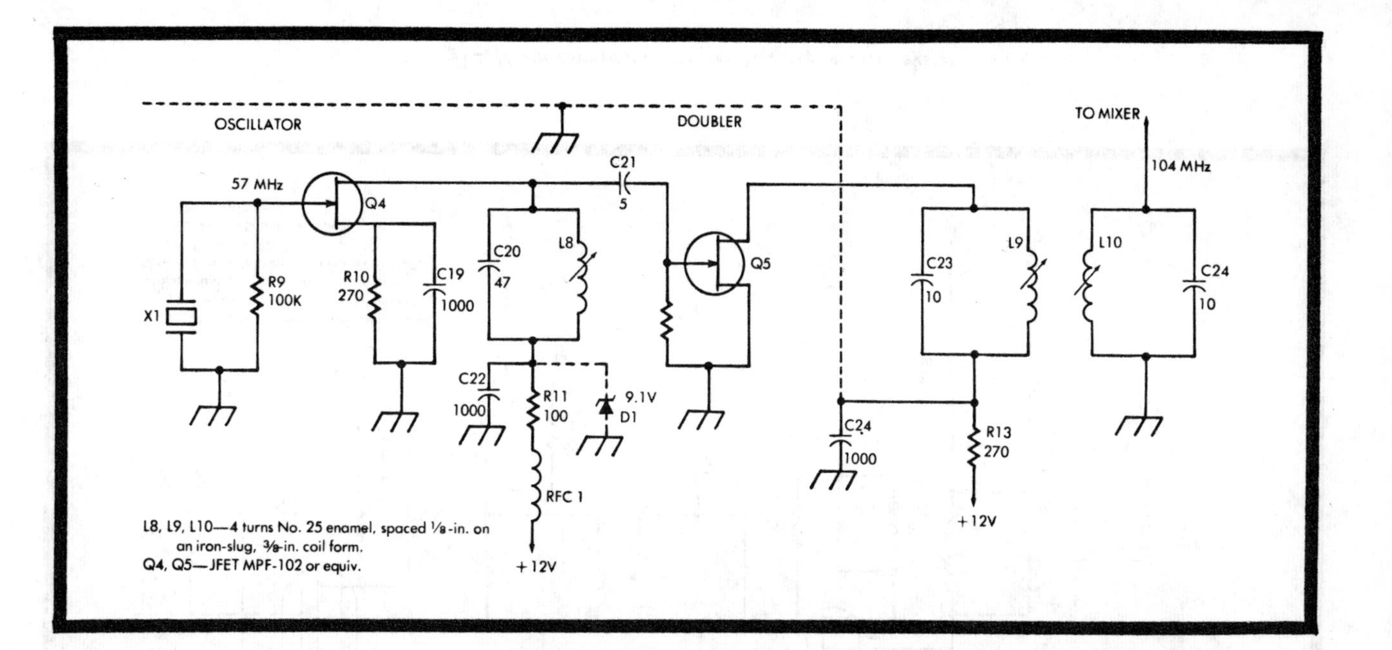

Fig. 9-11. Oscillator—doubler section for the 42 MHz system.

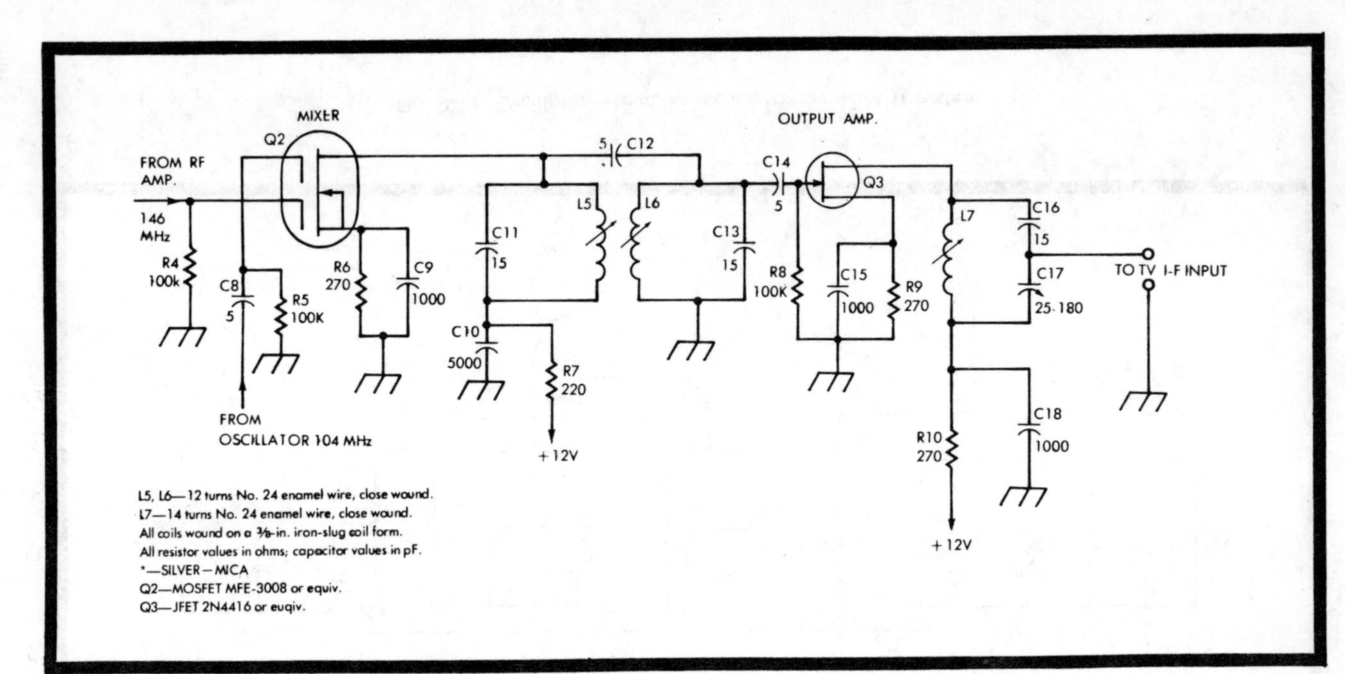

Fig. 9-12. Mixer and first i-f for the 42 MHz receiver system.

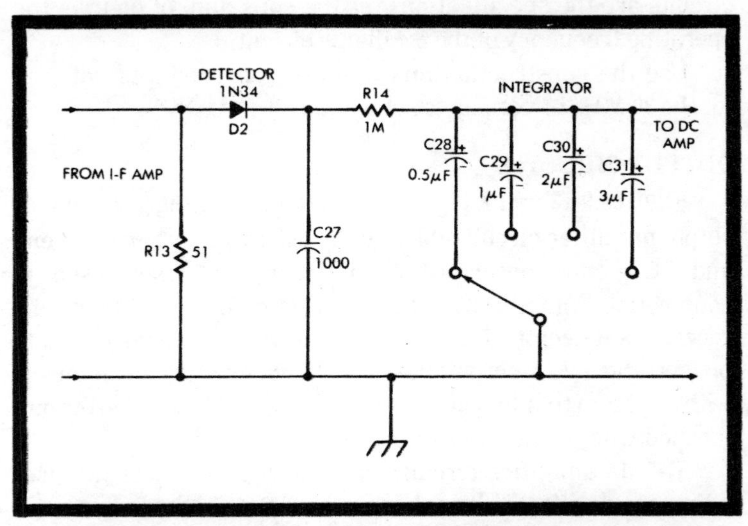

Fig. 9-13. Detector and integrator for the 146 MHz radio telescope systems.

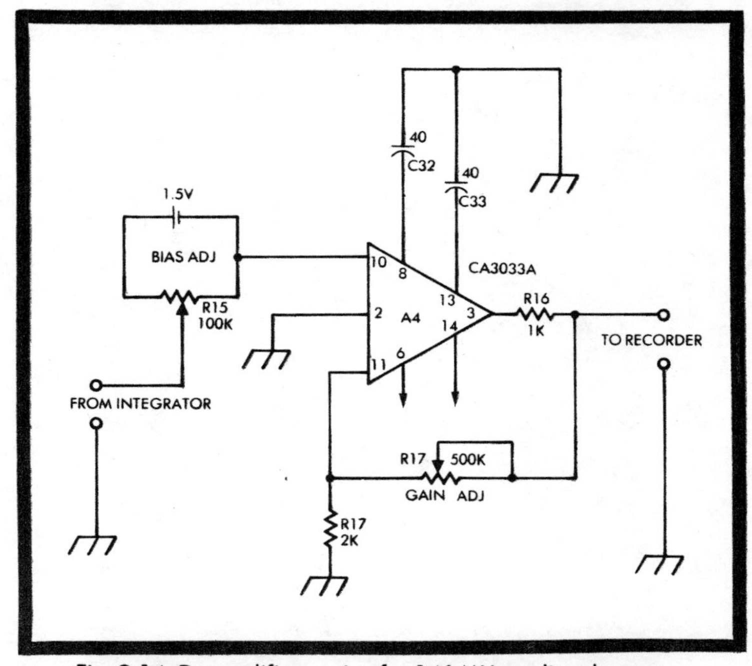

Fig. 9-14. Dc amplifier section for 146 MHz radio telescope.

circuits are the specifications for the coils and, of course, the operating frequency of the oscillator circuit.

Use the construction and alignment procedures outlined for the 30 MHz i-f version to set up this circuit.

OUTPUT CIRCUITS

Figures 9-13 and 9-14 show the detector–integrator and dc output amplifier circuits that can be used with either front end and i-f section described in this chapter. When used in conjunction with the 30 MHz circuits, diode D2 acts as the receiver's detector. The same is true when the user elects to pick off the TV i-f output ahead of the video stage. When used with a system that extracts its signal from a TV's video output, the diode merely serves as a signal rectifier.

The dc amplifier circuit shown in Fig. 9-14 lets the user adjust the dc offset voltage by means of R15. The other control, R17, acts as a gain control. The output of the IC amplifier can be fed to an oscillographic device or conventional meter.

Some Decameter Systems For Studying Jupiter

Much of the radio energy from the Sun and virtually all of the radio signals from Jupiter are of nonthermal origin. Nonthermal radiation is most distinct in the decameter range, and amateurs especially interested in studying nonthermal radio emissions from Jupiter will find the plans suggested here to be quite appropriate.

The basic system described in this chapter was suggested by Dirk Baker, a PhD candidate in electrical engineering at Ohio State University. Baker built and tested a decameter system for Jovian studies as part of his work for a master's degree.

GENERAL DESCRIPTION

The receiver system for decameter Jovian studies (at 18 MHz, for instance) is nearly always an ordinary shortwave receiver that is capable of tuning the 18–22 MHz portion of the shortwave band. The receiver should be of fairly good quality, however.

The receiver's usable sensitivity must be on the order of 1 μV or better. Anyone starting out with a high-performance receiver with a usable sensitivity of 0.5 μV is indeed getting off to a good start. A low-noise preamplifier is a necessary part of the receiver system whenever the receiver gain is not up to the 1 μV specification.

The good sensitivity of the receiver must be coupled with a high selectivity. The need for good selectivity might seem to run counter to the general rule that says all radio telescope receivers must have a wide bandwidth; but working Jupiter in

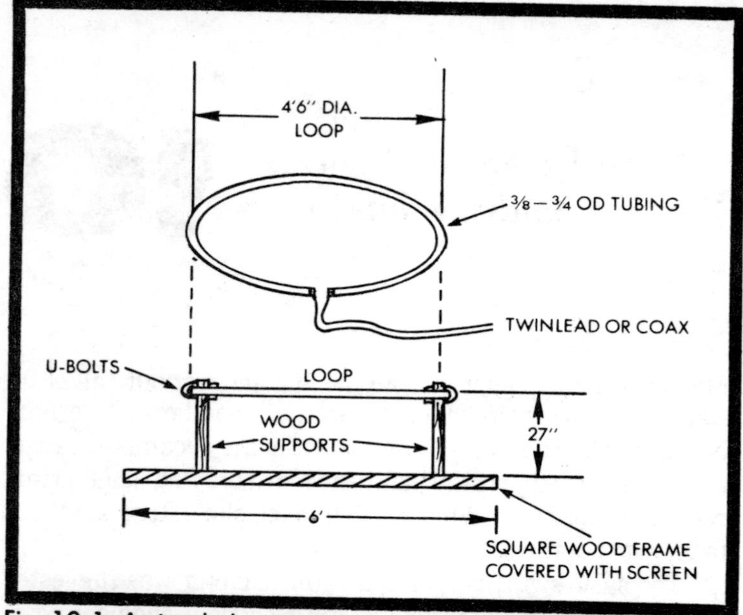

Fig. 10-1. A simple loop antenna for detecting decameter signals from Jupiter.

the decameter range is a different ballgame. The idea in this case is *not* to tune in sharply on Jupiter, but to tune sharply between the many terrestrial broadcasts that crowd the 18–22 MHz portion of the shortwave band. The whole receiver scheme relies upon the great power of Jupiter's radio transmissions to make up for a narrow receiver bandwidth.

After selecting a fairly good receiver—one having good sensitivity and selectivity—the amateur radio astronomer can turn his attention to the selection of a suitable antenna system.

The first antenna system described in this chapter is a very simple one that can be constructed and put into operation during a single weekend. The antenna is a simple loop antenna about 3½ ft in diameter. The scheme uses a wire-screen reflector to increase its front-to-back ratio.

The second antenna is a highly elaborate, high-performance 5-element beam that is cut for resonance at about 18 MHz. Whereas the loop antenna is more appropriate for beginners and casual observers, the beam is intended for serious amateurs and active radio research groups.

Regardless of the antenna system he uses, the experimenter can record the signals by tapping off the audio voltage from either the receiver's volume control or its headphone jack. Tapping the signal from the volume control allows him to fix the rf gain control at some convenient calibration level and yet adjust the audio gain so he can listen to the incoming signal if he wishes to do so.

The recording—readout system can be a voltmeter, chart recorder, or a combination of the two. Beginners often get a great deal of pleasure out of merely listening to the unusual "seashore" sounds of Jupiter through the receiver loudspeaker. Of course, the sounds can be recorded on one channel of a stereo tape recorder. If time signals are recorded on the other channel, the experimenter has a permanent record of Jovian radio transmissions that are correlated with the time of day.

SIMPLE 18 MHz ANTENNA SYSTEM

Figure 10-1 shows the electrical features of a simple loop antenna for detecting decameter signals from Jupiter (and the Sun, too, for that matter). The loop should be very close to one wavelength (1λ) in circumference for best performance; but at 18 MHz, the 1λ circumference would have to be on the order of 57 ft, which would make the diameter no less than 18 ft! A loop antenna of that dimensional scale destroys simplicity of the scheme. Even a half-wave loop antenna is bigger than most experimenters find practical—it would be about 9 ft in diameter.

To keep the system simple, it is necessary to reduce the circumference of the loop to ¼λ, or about 14 ft, 3 in. Using those dimensions, the diameter is about 4½ ft.

The frame for the reflector can be made from any conveniently available wood stock. The frame should be about 6 ft square to accommodate the 4½ ft loop antenna.

The loop itself can be made from any sort of metal tubing between ⅜ and ¾ in. in diameter. The loop should be mounted parallel to the reflector screen and 27 in. away from it.

Since the whole system relies upon large signal strength, a close impedance match between the antenna and transmission line is not a critical factor. Common TV twinlead will work, and so will 50–75Ω coaxial cable. An experimenter can match and balance the system to the extent that seem practical and desirable.

The loop antenna has a rather wide beamwidth that will allow it to pick up signals from Jupiter for about four hours out of every day. The antenna should be aimed at the local meridian (due south) and tilted upward toward the ecliptic—the path followed by the Sun and most of the planets. Whenever Jupiter's visible in the night sky, a casual observer can, of course, simply aim the antenna toward the planet and begin monitoring its signals.

INITIAL TESTS AND ADJUSTMENTS

When taking readings from the antenna or getting a noise-figure reading, set the rf gain control to maximum, turn off the AVC and squelch circuits, and set the audio gain control to a comfortable listening level.

To see if the system is working at all, disconnect the antenna lead from the receiver and replace it with a fixed resistor having an equivalent value (e.g., 50Ω, 75Ω, or 300Ω). The noise from the loudspeaker then represents the receiver's internal noise level. If a preamplifier is being used in the system, make sure the noise reading includes noise from that circuit as well. Record the S-meter reading or check the voltage from the desired signal pickoff point (at the volume control or headphone jack).

After getting a noise-level reading, remove the fixed load resistor from the antenna input terminals and replace it with the antenna lead. A sudden increase in noise level clearly indicates the system is working.

If there appears to be too little or no change at all in the noise level, it is most likely an indication that the system requires the help of a good preamplifier.

Determine the next transit time of Jupiter, and begin taking antenna readings at half-hour intervals about two hours before the transit is due. If there is any reason to believe the receiver's internal noise level is drifting, it is a good idea to check the noise level just before each antenna reading.

Continue taking antenna readings during the transit and for about two hours thereafter. Subtract the noise-level readings from the gathered data and plot the results on a sheet of graph paper. The graph should show an increase in antenna signal level that reaches a peak about the time of Jupiter's meridian crossing.

If a local broadcast should suddenly begin interfering with good reception of Jovian signals, it is advisable to change the receiver's frequency setting a little bit. Be sure to recheck the internal noise level after making the change, however.

Beginning experimenters should bear in mind the simple fact that Jupiter unaccountably falls quiet from time to time. If the system responds to general galactic background noise, evidenced by an increase in noise output when the receiver is connected to the antenna, the telescope is indeed working. An experimenter should not be disappointed if he finds nothing coming from Jupiter—just wait and listen to the signals for a couple of days.

SOME IMPROVEMENTS FOR ROUTINE OBSERVATIONS

When it is time to begin making observations of Jupiter on a routine basis, the experimenter should make a few refinements that simplify the job of gathering data and determining the receiver's inherent noise level.

It is possible to simplify the procedure for getting the system's noise figure by inserting a two-position antenna switch between the transmission line and receiver system. One of the positions should go directly to the antenna. The other positions can be filled with a dummy load. Taking receiver noise readings thus becomes a simple matter of changing the position of the selector switch.

The noise from the loudspeaker can become quite annoying after a time. Since most shortwave receivers have a headphone jack that automatically cuts out the loudspeaker when a plug is inserted into it, the experimenter can rig up a special plug that carries the audio output to a meter or other readout device. With the audio signal going directly to the output device, there is no need to listen to all the noise from the loudspeaker.

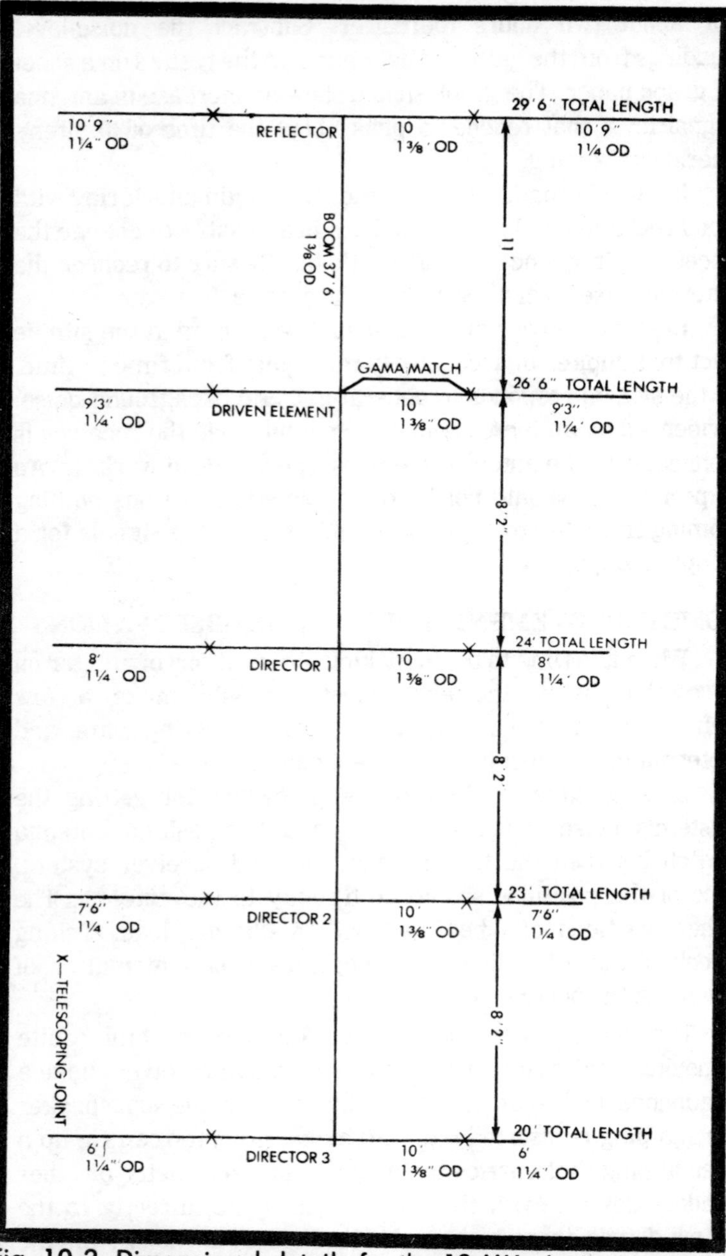

Fig. 10-2. Dimensional details for the 18 MHz Jupiter antenna. The X's mark the locations where tube sections are joined together.

There are some instances, however, where it is necessary to judge the quality of the audio output by listening to it. An abrupt increase in the recorder output, for example, might be due to a local transmission rather than some fantastic phenomenon taking place on Jupiter. Unplugging the recorder and listening to the audio signal for a moment is usually enough to clear up any question about the origin of the signal.

HIGH-PERFORMANCE ANTENNA SYSTEM

The antenna and its mounting are the only major parts of this particular radio telescope system that are really worthy of being built up from scratch. Of course, 15-meter antennas are available through amateur radio shops, but few have the same high-performance characteristics as the one described here. None of the commercial versions are intended to be used with the required 20° tilt from the horizontal.

Assembling the Beam

Figure 10-2 shows the main construction features of the 5-element beam antenna. The boom and middle sections of the elements are made of 1⅜ in. steel tubing. Chain link fence tubing is usually available in cut-to-order lengths up to 21 ft. Since the boom is about 36½ ft long, it requires one 21 ft section joined to a section 15½ ft long. The two sections can be joined with a top rail sleeve that is bolted to the sections for added strength.

The center sections of all five elements are cut to 10 ft and joined to the boom with muffler clamps as suggested in Fig. 10-3. The overall lengths of the elements are adjusted by means of telescoping sections of smaller diameter electrical conduit. The conduit in this instance should have an outer diameter of 1¼ in. Install the pieces as shown in Fig. 10-3. The dimensions shown in Fig. 10-2 assume the 1¼ in. sections are inserted about 1 ft into the larger tube.

The antenna uses two sections of 1¼ in. electrical conduit as support members that eliminate drooping at the ends of the boom. Since the conduit is supplied in standard 10 ft lengths, the forward section has to be made up of one full length having a 5½ ft length joined to it. (See Fig. 10-6.) Join the two support pieces with a 1⅜ in. OD sleeve. Drill a ¼ in. hole about 1 in.

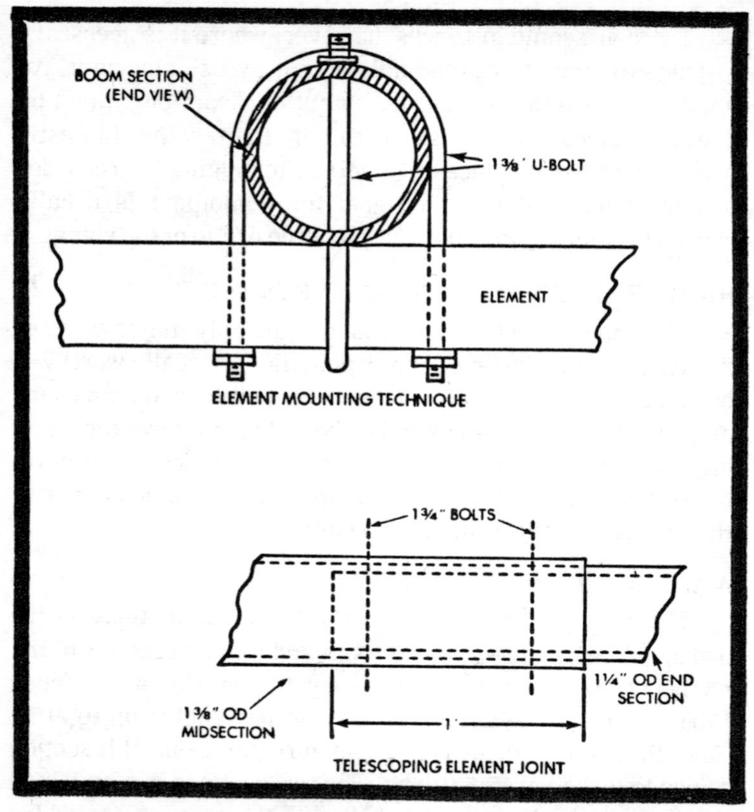

Fig. 10-3. Element mounting and telescoping detail.

from each end of both support pieces. Drill two ¼ in. holes through the boom for attaching the support members, but do not make any connections at this time.

After completing the assembly of the boom and elements locate the center of gravity by lifting the boom near its center. Since the assembly weighs about 100 lb at this point in construction, it is advisable to enlist the aid of a burly assistant. Mark the balance point on the boom and drill the mast mounting holes as shown in Fig. 10-5.

Construct a *gamma* match assembly (consult reference book or amateur radio handbook), raise the boom about 8 in. from the ground at the driven element, and temporarily install the matching assembly to check its fit. When the gamma assembly appears to be in good order, remove it until the antenna is partly erected. (The matching section is likely to be damaged during the setup procedure.)

Assembling and Erecting the Mast

The mast is made of ordinary pine 2×4 stock that has been cut to size and coated with several layers of spar varnish or deck enamel. (See Fig. 10-6 for construction details.)

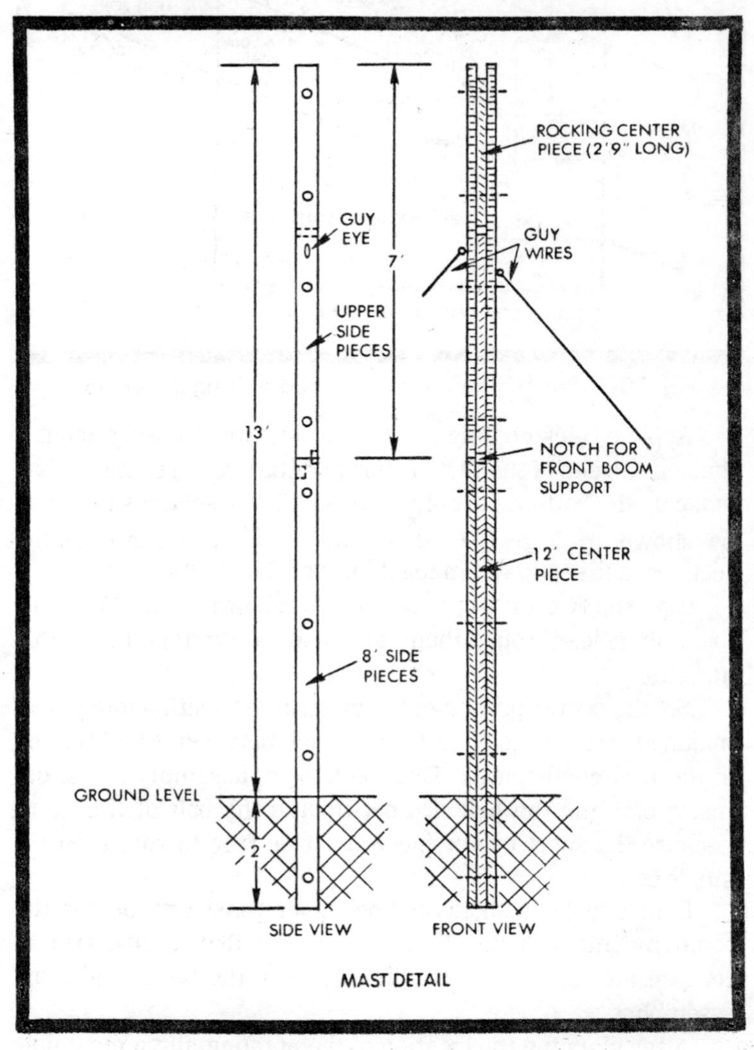

Fig. 10-4. Mast construction details.

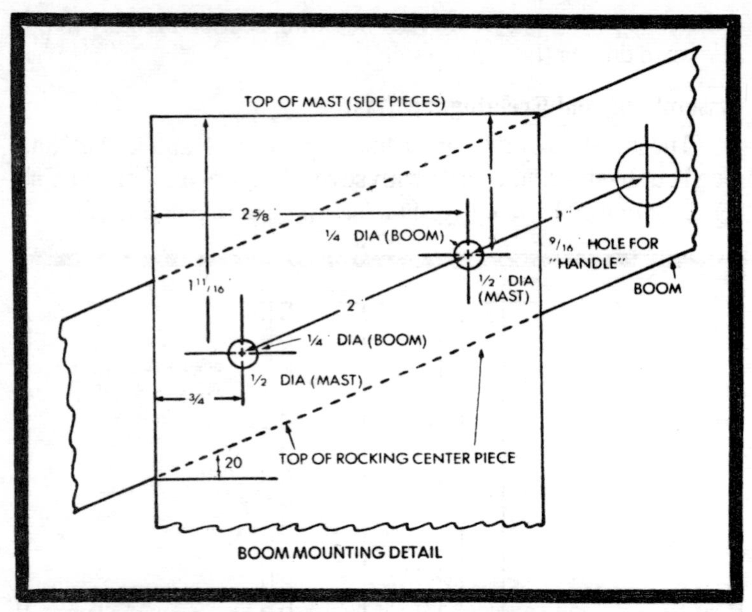

Fig. 10-5. Boom mounting detail and drilling dimensions.

Apply a thick coating of pitch or blacktop dressing about 2 ft along one end of the 12 ft section and the two 8 ft sections that make up the bottom part of the mast. Fit the sections together as shown in Fig. 10-6, drill four ⅜ in. holes through the sections, and secure them together with 6 in. bolts.

Align the two 7 ft upper sections as shown in Fig. 10-6, drill the ⅜ in. holes through them, and fasten them together with 6 in. bolts.

Set the center portion of the upper mast section into place, making certain to leave a 1½ in. space between it and the top of the 12 ft centerpiece. Drill both ⅜ in. assembly holes, but insert only the top bolt. Do not tighten the bolt at this time, because the short center piece must be free to rotate on the single bolt.

Drill two ½ in. holes in the upper mast section for the boom mounting bolts. These holes are drilled oversize to compensate for slight misadjustments in the boom and mast assemblies.

After planning the location of the antenna, dig a mast hole about 8 in. across and a bit over 2 ft deep. Spread about 3 in. of dry sand, gravel, or fine stones in the bottom of the hole, and drop the mast into it. Fill the hole about ¾ full of a concrete mixture and align the mast so that its front section faces south. Adjust the mast to true vertical with the aid of a level or plumb bob. Then fill the hole to the top with the cement mixture, realigning the mast if necessary.

Erecting the Beam

When the concrete hardens, arrange the beam section flat on the ground so that it is lying straight north and south. Butt the director end of the beam against the north side of the mast.

Set up a 5 ft stepladder on the south side of the mast and lash it to the mast assembly. Rotate the movable inner section of the upper mast so that its top points southward.

Raise the south end of the beam, and work the tip around the movable upper mast section and into the slot between the outer sections of the upper mast.

While a helper on the ground slides the boom forward and upward, raise the boom so that the first director element

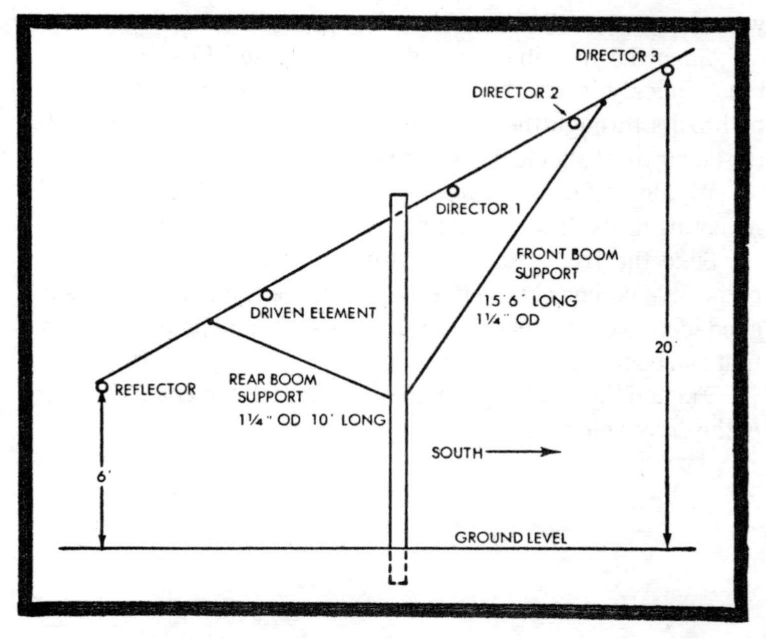

Fig. 10-6. Side view of the finished antenna—guy wires, gamma match, and transmission line not shown.

clears the mast. Continue feeding the boom through the slot in the upper mast section until the second director element is raised over the south side of the mast. Slide the boom about 6 in. farther, and let the north end of boom come to rest on the ground.

Bolt the forward (15 ft) boom support to the west side of the boom and fix the two uppermost guy wires in place. Let the mast end of the boom support and the guy wires hang free. Bolt the shorter boom support to the west side of the lower end of the boom, fasten the two lower guy wires into place, and allow the unattached ends of these components to hang free. Fix the gamma match assembly to the driven element.

Resume feeding the beam through the upper mast section until the first director element is raised over the mast. Inch the beam forward until the boom mounting holes are aligned with the mounting holes in the upper mast section. Grasp the lower end of the movable mast section and pull it toward the mast. This is the most difficult portion of the procedure, because the boom seats itself with a panicky sort of jolt. It is wise to exercise all reasonable safety precautions at this point.

When the movable mast section is pulled flush with the mast, quickly insert the second mast bolt into place. Tighten both bolts through the upper mast section. The inner part of the upper mast should no longer be movable.

Wiggle the beam assembly to realign the boom-to-mast mounting holes. Insert and tighten both of the mounting bolts.

Slide the free ends of the two boom supports into their respective notches in the mast and bolt them into place. It is a good idea to double-check the tightness of all bolts and screws that can be reached from the stepladder.

Fix and tighten the guy wires, and attach the coaxial cable to the driven element.

A Complete Solar Radio Observatory

11

The Sun has been the subject of more prolonged and intensive radio studies than any other single object in the sky. Radio studies of the Sun are valuable to astronomers for a number of reasons; but the work is especially meaningful when it concerns solar flare activity.

Optical observation of flare activity has always been hampered by the simple fact that it is very difficult to study highly selective portions of the Sun's bright surface with ordinary optical equipment. In fact, most optical studies of flare activity can be carried out only during total eclipses, when the Sun's corona is all that is visible. Few amateur optical astronomers have ever had an opportunity to observe a solar flare directly.

Figure 11-1 shows a photo of a solar flare. Although the flare appears to be a few simple tentacles of solar matter reaching into space, it actually represents the most intensive energy exchange that ever takes place in our part of the galaxy.

Solar flares create immense doses of radio energy that seem to be most intense in the 137 MHz range. A radio astronomer tuning that particular frequency range can detect a growing flare about eight minutes after its activity reaches a critical point. The eight-minute delay is, of course, due to the travel time of radio waves between Sun and Earth.

A solar flare also emits ultraviolet rays and a host of subatomic particles that produce a secondary effect of special interest to radio astronomers: The rays upset the normal daytime characteristics of the Earth's upper atmosphere. Amateur radio operators are well aware of the fact that the Earth's upper atmosphere "settles down" during the night. The occurrence of a solar flare mimics this effect, making terrestrial broadcasting especially good during daytime hours.

This chapter describes the equipment necessary for monitoring direct solar activity at 137 MHz and sudden enhancement of radio propagation at the VLF spectrum position of 27 kHz.

137 MHz SOLAR RADIO TELESCOPE

The 137 MHz radio telescope system described here is made up of a corner-reflector antenna and a receiver system that includes either a 108–135 MHz aviation band receiver or a combination 137 MHz converter and amateur shortwave receiver.

The Antenna

Figure 11-2 shows the critical dimensions and the main construction features of the 137 MHz corner reflector. The driven element is a simple half-wave dipole that is matched to the transmission line by means of a standard gamma arrangement. The gamma match allows the experimenter to use a low-noise coaxial transmission line, which is less troublesome than the *balanced line* that might otherwise be required.

The antenna mounting shown in Fig. 11-2 does not permit easy adjustment of the antenna's azimuth setting, but the altitude can be adjusted by swinging the rear supports toward or away from the reflector frame. For all practical experimental purposes, it is better to leave the antenna pointing toward the meridian, anyway. The altitude can be adjusted slightly from time to time to compensate for seasonal changes in the Sun's altitude as it crosses the local meridian.

The corner-reflector arrangement allows a very wide beamwidth in the east–west direction—half-power points are on the order of 120° apart. An experimenter can thus expect to

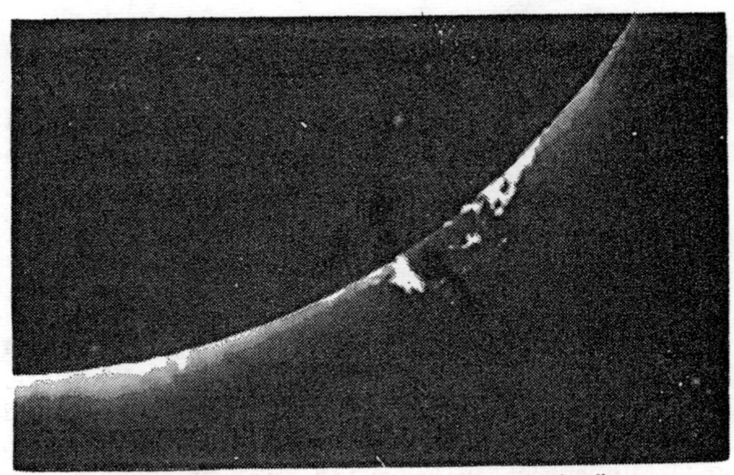

Fig. 11-1. Lick Observatory photo of a typical solar flare.

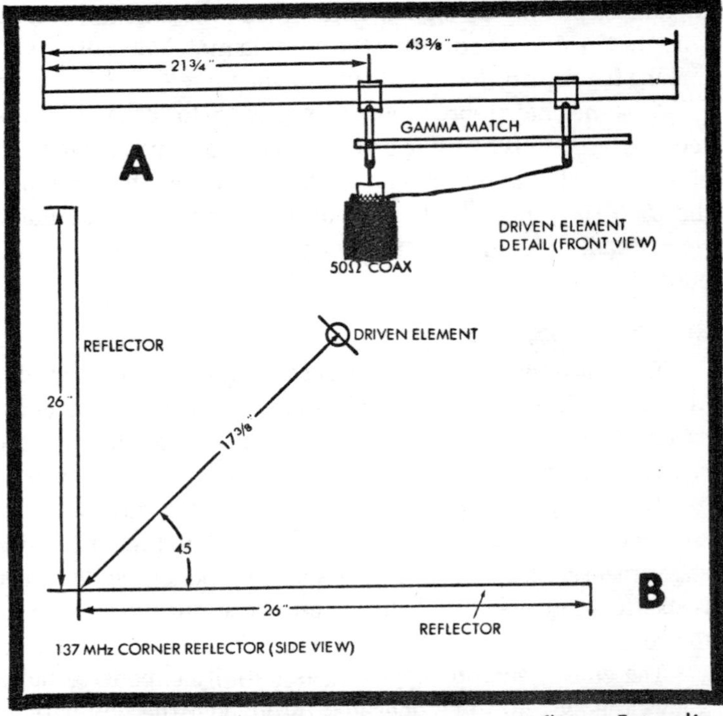

Fig. 11-2. Dimension detail for the 137 MHz corner reflector. Extending reflector length beyond 26 in. would increase gain somewhat. At 88 in. the gain would be about 10 dB.

pick up signals from the Sun for about eight hours out of every day.

The gain of the antenna is very low compared to that of other types of high-performance antennas used in radio astronomy work. This low gain factor poses no real problem, however, because solar flare radio emissions are very intense at 137 MHz.

137 MHz Receiver Systems

The simplest 137 MHz receiver system consists of an aviation band receiver that tunes the 108–135 MHz range. Most receivers allow a bit of frequency overrun at either end of the band, so the receiver will most likely be able to tune up to 137 MHz without your having to do any interal tampering. If a check with an rf signal generator shows the receiver is *not* capable of picking up signals at 137 MHz, the antenna can be tuned to 135 MHz by adjusting the gamma match or, better yet, making the driven element about 4 in. longer.

A somewhat more roundabout way to assemble the receiver system is by using an aviation-band converter in front of a standard amateur-band receiver that is capable of tuning the 30 MHz range. Such aviation converters are available ready-made and in kit form from a number of electronics outlets.

Recording Systems

There are two basic kinds of recording systems for the 137 MHz solar radio telescope: a simple voltmeter or a chart recorder. In either case, the output can be taken from the receiver's earphone jack or its volume control connections.

The voltmeter approach has the distinct advantage of simplicity and relatively low cost. The disadvantage is that the experimenter has to be on hand to spot a solar flare response—and that can occur at any time during any day of the year.

The chart recorder approach is initially expensive but it allows the system to be operated throughout the day without any special attention. All the experimenter has to do is load fresh chart paper, note the time, and let the system run; and at the end of the day, he scans the chart for a flare response that will always appear as a marked change in signal output.

A third kind of recording system strikes a compromise between the advantages and disadvantages of both the voltmeter and chart recorder. The experimenter can connect the audio output of the telescope receiver to one tape track of a stereo tape recorder. The audio output from a second receiver that is tuned to a time-standard station such as WWV or CHU can go to the second recording track. The result is a two-track recording having all the raw signal data on one channel and time signals on the other. The experimenter can play back the recording, monitoring the 137 MHz radio signal by means of a voltmeter. Whenever he notes a sudden and large increase in signal level, he can read the amount of signal change on the meter and manually record the time of occurrence from the time channel.

The tape recorder technique thus allows the system to operate unattended for long periods of time, and yet does not require an expensive chart recorder. Of course it can be argued that a two-channel tape recorder is perhaps more expensive than a chart recorder; but it also must be argued that the tape recorder has far more applications outside the realm of radio astronomy than a specialized chart recorder does.

Standard Operating Procedures

Align the antenna so that the long, open end of the corner reflector is pointed toward the meridian (due south). Adjust the altitude so that the antenna points to the place in the sky where the Sun crosses the meridian.

Turn on the receiver system; set the rf gain control to maximum; turn off the AVC, squelch, and BFO controls; and set the audio control for a comfortable listening level.

Disconnect the transmission line from the receiver and replace the leads with a fixed resistor having a value equal to the characteristic impedance of the transmission line. Record the noise level from the receiver. That recorded figure represents the system's internal noise level, and should be subtracted from all subsequent readings from the antenna.

Replace the transmission line connection. If the system is working properly, the recording device should register at least a very slight increase in signal level.

Record the receiver's output signal as the Sun begins passing through the antenna beam aperture. It is a good idea to recheck the system's noise level at regular intervals throughout a recording session.

27 kHz S.E.A. MONITORS

The thunderstorm activity normally taking place at hundreds of points on Earth at any given instant generates a constant stream of static that seems to be most intense in the VLF range—at about 27 kHz. As the Earth's upper atmosphere adjusts itself to favor long-range radio communications, these normal sources of atmospheric noise tend to become more intense as reproduced by a VLF receiver.

The enhancement of atmospheric noise usually takes place every day between sundown and sunrise. Whenever a solar flare occurs, however, the *sudden enhancement of atmospherics* (S.E.A.) phenomena can take place during the daylight hours as well.

Recording radio signals at 27 kHz during the daylight hours is thus one very effective means of monitoring the occurrence of solar flares. Whenever the radio signal shows a sudden increase—a jump toward the high nighttime levels—the experimenter can be reasonably certain he is observing the effects of a solar flare.

The equipment for S.E.A. monitoring is rather simple and inexpensive. An experienced technician can have a system set up and running within a couple of days.

The Antenna System

The antenna system for S.E.A. recording at 27 kHz can be a simple long-wire antenna or a special type of vertical antenna suggested by the *Solar Division* of the AAVSO.

The orientation of a long-wire antenna for S.E.A. work is not very critical, although running the wire from north to south might produce somewhat better results. The wire should be at least 100 ft long—the longer the better. If space permits, a horizontal rhombic measuring about 50 ft to a side works quite well, too.

The AAVSO antenna is made up of two 10 ft sections of 3 in. downspout tubing. The downspout is mounted vertically on top of a 3 ft section of 2-by-4 lumber that is sunk into the ground and trimmed on the exposed end to slide into the downspout tubing.

The signal pickoff point on the vertical antenna is at the joint for the two sections of tubing. There should be no attempt to match impedances anywhere in this particular system—it relies upon brute-force signal strength to overcome any attenuation due to impedance mismatches. It is important, however, to take advantage of the unbalanced nature of the antenna, and use low-noise coaxial transmission line to carry the signal from the antenna to the receiver system.

Fasten the center conductor of the coax to a clean, tight connection at the center of the antenna. Make the outside conductor connection at the ground terminal on the receiver.

Receiver System

Receivers for the VLF bands are relatively easy to design, build, and align. The technology is much like that of audio circuitry, making component selection and layout much less critical than it is for any other kind of receiver circuit.

A very simple 27 kHz receiver system appears in the "Amateur Scientist" department of *Scientific American* magazine for September 1960. A somewhat more up-to-date and elaborate system appears in the June 1973 issue of *Sky and Telescope*. Interested experimenters should consult these two articles for more details about S.E.A. monitoring as well as for the radio schematics they contain.

Normal Operating Procedures

The biggest operational problem associated with 27 kHz S.E.A. monitors is that of distinguishing Sun signals from local man-made electrical noise. Man-made sources of 27 kHz radio noise nearly always come from some sort of spark-discharge apparatus such as unshielded auto spark plugs, certain types of fluorescent lighting, and motors having a bad set of electrical brushes.

With some experience, an experimenter can learn to tell the difference between a legitimate S.E.A. and a man-made source by observing the quality and time scale of the signal. If, for example, the signal seems to be switching off and on at a fairly regular rate, chances are quite good the noise is man-made. Likewise, signals appearing and disappearing within a matter of minutes are probably of local origin.

The surest way to tell the difference between an S.E.A. and local man-made noise is to set up two or more receiving stations at least five miles apart—the greater the separation, the better. Any signal appearing at the same time on all the monitors has to be either an S.E.A. or the effect of a widespread thunderstorm. Since it is pointless to try geting anything meaningful out of recordings made during a local thunderstorm, the multimonitor scheme turns out to be the most reliable in all respects.

The multimonitor technique for S.E.A. observation, by the way, can be made especially fascinating and meaningful if the widely separated observers are qualified to set up an amateur radio link between their monitoring stations.

The Amateur's Challenges

One of the recurring themes of this book is that radio astronomy, both on professional and amateur levels, is still in its infancy. This is especially true in the case of amateur radio astronomy, where equipment has to be built from scratch or patched together from items intended for other purposes.

The amateur experimenter's immediate challenges are to: (1) develop specialized equipment and circuits that are useful, practical, and easy to use; and (2) establish some standard operating procedures that promise satisfying and meaningful results.

The first challenge—coming up with some specialized circuits—is mainly an engineering one. Most of the projects suggested in this book represent the first steps toward developing practical and useful radio telescope systems; and the primary objective has been to show beginners how to assemble workable systems from inexpensive parts and equipment that are readily available from electronics shops and hardware stores. There are still plenty of opportunities for developing new kinds of circuits and equipment—the job, in fact, has just barely begun.

There are several good reasons why it is now important to develop some standardized operating procedures for amateur radio astronomy. On one level, the idea is to attract newcomers and maintain their interest long enough to make them into active enthusiasts. The best way to attract newcomers is to offer them opportunities to construct equipment that is different from anything else around and give them a chance to make some exciting observations that are likewise unique. A newcomer working without the benefit of the experiences of others, however, is bound to run into difficulties; and the only way to maintain his interest is by insuring some satisfying results by means of a few tried-and-true experiments.

Another reason for setting up standard operating procedures is to make it easier for amateur radio astronomers to exchange ideas and compare results. Real progress in amateur radio astronomy will rely upon the cumulative efforts of many individuals. If every experimenter works in a different direction and according to different sets of standards, it will be difficult to see any progress at all. Of course, any amateur should be able to strike out on his own with original experiments, but there has to be some basic standards that serve as a meaningful point of departure.

The following sections deal primarily with the technical problems now facing experimenters who would like to see amateur radio astronomy move ahead as a legitimate technical hobby: finding ways to take advantage of the vast UHF region of the spectrum and developing radio telescope systems having far better resolution and sensitivity than is possible with ready-made equipment now available at reasonable costs.

THE UHF GAP

All of the systems described in this book operate in the VHF part of the radio spectrum or below. The reason is simple: UHF equipment having the necessary specifications is simply not available on an off-the-shelf basis. If the right kind of UHF equipment *were* available, projects working in that part of the spectrum would fill a better share of this book.

The UHF region occupies most of the useful radio astronomy band, but there are several far more compelling reasons for wanting to work at centimeter wavelengths. For one thing, stellar radiation of thermal origin is more active at UHF. Then, too, the UHF bands are relatively free of natural and man-made interference. And finally there is a matter of practical significance—UHF antennas are far smaller and less expensive than their VHF counterparts of comparable gain.

It is tempting to try using a commercial television receiver equipped with a UHF tuner to work the 470–890 MHz range. Don't bother. Unless an experimenter is willing to completely revamp the tuner and couple it to a high-gain rf preamplifier, there is little point in considering a TV receiver at all.

The trouble with ordinary television receivers is their intolerably poor sensitivity. Most sets around today have sensitivities on the order of 200 μV or worse. Even newer sets that have MOSFET front ends do not begin to approach the 5 μV sensitivity of an ordinary FM receiver.

One might argue that the extraordinarily wide predetection bandwidth of a TV set makes up for its poor sensitivity. That is not the case, however. As demonstrated by the basic equations in Chapter 5, the power available at the antenna is indeed proportional to the system bandwidth. An experimenter would have a fine radio telescope at his disposal if he could take advantage of a video bandwidth of 3 MHz. The same set of equations, however, show that the minimum power that must be available at the receiver's input increases with the square of the sensitivity figure. Using a TV receiver in place of an FM receiver increases the system bandwidth and effective antenna power 30 times over. That's nice. But the corresponding drop in sensitivity from 5 μV to about 200 μV means increasing the power that must be available at the receiver 1600 times!

Working with the equations in Chapter 5 shows that an ordinary UHF television receiver can detect signals from the Sun and Jupiter, providing the antenna has a gain of about 30 dB or better. Although the individual antenna elements at 800 MHz would be rather small, the antenna array necessary to achieve a gain of 30 dB, even with the help of a standard rf preamplifier, would approach the cost and complexity of a lower-gain beam antenna cut for VHF. And besides, the nonthermal emissions of the Sun and Jupiter are more intense and interesting in the decameter range.

Several amateurs have devised low-noise front ends for television receivers; but by the time they go to all that trouble, they might well have added another stage or two to replace the TV set altogether.

It is a frustrating situation. TV sets have an immense bandwidth that is most appealing to radio astronomy work. Unless some manufacturer decides to put out a set that has a sensitivity on the order of 10 μV or better (and no one has any real reason to do that), there is no practical way to take advantage of all that bandwidth potential.

After carefully considering the notion of using a TV set for working in the upper VHF and UHF portions of the radio astronomy spectrum, it might occur to an experimenter that he could couple a UHF converter to the input of an FM tuner. The idea is a sound one in principle: take advantage of the good sensitivity figures of an FM tuner and use a wideband UHF converter to reach into the UHF part of the spectrum. It also helps to take advantage of the fact that off-the-shelf television UHF converters generate a conversion frequency close to that of VHF channel 2, and it is not difficult to adjust that frequency into the FM band.

Unfortunately, the converter/FM trick isn't practical. The main trouble is that the signal power at the antenna terminals drops off as the effective aperture of the antenna decreases (see Chapter 5). The antenna elements are indeed smaller at UHF, but a large array is needed to boost the effective aperture to that of a single beam antenna at VHF. In short, the size and complexity of the necessary antenna array outweighs any advantage gained by using a UHF converter coupled to a good FM receiver.

What, then, is needed to work the vast UHF region of the radio astronomy spectrum? The solution would be a high-sensitivity receiver with a very wide bandwidth. At an operating frequency of 800 MHz, for example, a receiver with a sensitivity of 1 μV and a bandwidth on the order of 3 MHz could reliably detect 1000 F.U. (flux unit) radio sources using an antenna with 20 dB of gain. The antenna would not have to be very complicated—a small 4-bay helical array measuring about 2 ft square could do the job. The problem is to get both sensitivity and high operating frequencies in the same receiver circuit.

RESOLUTION AND SENSITIVITY

Amateur optical astronomy has been a popular pastime for generations. One of the primary features of optical astronomy that attracts and sustains the interest of so many people is the fact that an observer can use a relatively inexpensive telescope to view an almost unlimited variety of celestial objects. This is not the case with amateur radio astronomy today.

Using equipment that is readily available, an amateur radio astronomer can expect to observe no more than four to six different radio sources. To be sure, some of the extragalactic sources are unobservable with amateur optical equipment. It is doubtful, for example, that a viewing observer will ever see Cygnus A with a modest optical system (Fig. 12-1), but the radio observer will find it to be the most powerful source of extragalactic signals yet discovered. Still, it is difficult to interest newcomers in an activity that promises so little in the way of day-to-day rewards.

Serious amateur radio astronomers at work today belong to a special breed of experimenters. For one thing, they were probably attracted by the challenge of assembling some unique communications systems. And once they got the systems working, projects of the kind outlined in Chapter 6 sustained their interest. Most people already working with communications electronics on a professional or amateur level can feel the challenges of assembling a radio telescope; but it takes a peculiar kind of individual to put up with the tedious job of doing some useful work with the limited variety of radio sources.

Sooner or later, serious amateurs are going to have to face the problem of engineering some reliable, sensitive, and relatively low-cost interferometer systems capable of resolving less than 1° of arc at flux-unit levels of 100 or less. Only through the development of such systems can amateur radio astronomy spark and maintain the interest of newcomers the way optical astronomy does.

The path leading from the present-day situation to one that is more attractive in terms of variety and excitement is clear. It all goes back to the problem of developing a wideband UHF system with a sensitivity of 1 μV or better.

A radio telescope could resolve close to 1° of arc at 100 F.U. if the predetection bandwidth is 10 MHz, the receiver sensitivity is on the order of 1 μV, and the system is operated in the interferometer mode at 800 MHz. The antenna specifications would not be very rigid—20 dB each would be adequate if they are spaced at least 20 wavelengths apart.

It is impossible to say at this time how well amateur radio astronomy will progress as a popular pastime. With the right kind of organization and technical discipline, the next generation of young experimenters might be able to take radio astronomy projects for granted.

APPENDIXES

Appendix A
A Listing of Brighter Reference Stars

Astronomical Name and Abbreviation		Common Name	Magnitude	R.A. h	m	dec. (degrees)
αEridani	αERI	Achernar	1	1	37	−57
αUrsae Min.	αUMI	Polaris	2	2	06	+89
αArietis	αAIR		2	2	06	+23
oCeti	oCET		2	2	18	−03
βPersei	βPER		2	3	22	+50
αTauri	αTAU	Aldebaran	0	4	34	+16
βOrinis	βORI	Rigel	0	5	13	−08
αAurigae	αAUR	Capella	0	5	15	+46
γOrinis	γORI	Bellatrix	2	5	24	+06
βTauri	βTAU	El Nath	0	5	25	+29
ϵOrinis	δORI		2	5	35	−01
ζOrinis	ζORI		0	5	40	−02
αOrinis	αORI	Betelgeuse	0	5	54	+07
βAurigae	βAUR		2	5	58	+45
βCanis Maj.	βCMA		2	6	22	−18
αCarinae	αCAR	Canopus	−1	6	23	−53
γGeminorum	γGEM		2	6	36	+16
αCanis Maj.	αCMA	Sirius	−1	6	44	−17
ϵCanis Maj.	ϵCMA		2	6	58	−29
δCanis Maj.	δCMA		2	7	07	−26
αGeminorum	αGEM	Castor	2	7	33	+32
αCanis Min.	αCMI	Procyon	0	7	38	+05
βGeminorum	βGEM	Pollux	0	7	44	+20
γVelorum	γVEL		2	8	09	−47
ϵCarinae	ϵCAR		2	8	22	−59

Astronomical Name and Abbreviation		Common Name	Magnitude	R.A. h	m	dec. (degrees)
δVelorum	δVEL		2	8	44	−55
βCarinae	βCAR		2	9	13	−70
αHydrae	αHYD		2	9	26	−09
αLeonis	αLEO	Regulus	2	10	07	+12
γLeonis	γLEO		2	10	19	+20
αUrsae Maj.	αUMA	Dubhe	2	11	02	+62
αCrucis	αCRU		1	12	25	−63
γCrucis	γCRU		2	12	30	−57
αCrucis	βCRU		1	12	46	−60
ϵUrsae Maj.	ϵUMA	Alioth	2	12	53	+56
αVirginis	αVIR	Spica	1	13	24	−11
ηUrsae Maj.	ηUMA	Alkaid	2	13	47	+49
βCentauri	βCEN		1	14	02	−60
αBootis	αBOO	Arcturus	0	14	14	+19
αCentauri	αCEN		0	14	38	−61
αScorpii	αSCO	Antares	1	16	28	−26
αTrianguli Australis	αTRA		2	16	46	−69
γScorpii	γSCO		2	17	32	−37
θScorpii	θSCO		2	17	35	−43
ϵSagittarii	ϵSAG		2	18	22	−34
αLyrae	αLYR	Vega	0	18	36	+39
αAquilae	αAQU	Altair	1	19	50	+09
αPavonis	αPAV		2	20	24	−57
αCygni	αCYG	Deneb	1	20	41	+45
αGruis	αGRU		2	22	07	−47
αPiscis Austrinis	αPSA	Formalhaut	1	22	56	−30

Appendix B

Radio Sources of Interest to Amateurs

Name	Signal Strength (flux units)	Right Ascension hr	min	Declination
Taurus A	1420	05	30	+22°
Virgo A	970	12	28	+13°
Sagittarius A	9000	18	00	−24°
Cygnus A	8100	19	58	+41°
Cassiopeia A	11,000	23	21	+59°
Jupiter	10^5 (typical)	coordinates vary		
Sun	10^2 to 10^{10}	coordinates vary		

Appendix C
Sidereal Timetable

(0^h Universal Time)

Date	Sidereal Time (decimal hours)
JAN 3	7.06
7	7.32
11	7.58
15	7.84
19	8.12
23	8.37
27	8.63
31	8.89
FEB 4	8.96
8	9.24
12	9.50
16	9.76
20	10.03
24	10.29
28	10.55
MAR 3	10.82
7	11.08
11	11.34
15	11.60
19	11.87
23	12.13
27	12.39
31	12.66
APR 4	12.92
8	13.18
12	13.45
16	13.71
20	13.97
24	14.23
28	14.50

SIDEREAL TIMETABLE)

(0^h Universal Time)

Date	Sidereal Time (decimal hours)
MAY 2	14.76
6	15.02
10	15.28
14	15.55
18	15.81
22	16.07
26	16.34
30	16.60
JUN 3	16.86
7	17.12
11	17.39
15	17.65
19	17.91
23	18.18
27	18.44
JUL 1	18.70
5	18.96
9	19.23
13	19.49
17	19.75
21	20.02
25	20.28
29	20.54
AUG 2	20.80
6	21.07
10	21.33
14	21.59
18	21.86
22	22.12
26	22.38
30	22.64

SIDEREAL TIMETABLE
(0^h Universal Time)

Date	Sidereal Time (decimal hours)
SEP 3	22.91
7	23.17
11	23.43
15	23.70
19	23.96
23	00.22
27	00.48
OCT 1	00.75
5	01.01
9	01.27
13	01.54
17	01.80
21	02.06
25	02.32
29	02.59
NOV 2	02.85
6	03.11
10	03.38
14	03.64
18	03.90
22	04.16
26	04.43
30	04.69
DEC 4	04.95
8	05.22
12	05.48
16	05.74
20	06.00
24	06.27
28	06.53

NOTE: These sidereal times are accurate to within $\pm 0.25^h$. The figures vary somewhat on a 4-year cycle. See the *American Ephemeris and Nautical Almanac* for the current year to determine the exact sidereal times for any day.

Appendix D

Three-Place Sine and Cosine Tables

A	sinA	cosA	A	sinA	cosA
0	.000	1.00	50	.766	.643
2	.035	.999	52	.788	.616
4	.070	.998	54	.809	.588
6	.104	.994	56	.829	.559
8	.139	.990	58	.848	.530
10	.174	.985	60	.866	.500
12	.208	.978	62	.883	.470
14	.242	.970	64	.899	.438
16	.276	.961	66	.914	.407
18	.309	.951	68	.927	.375
20	.342	.940	70	.940	.342
22	.375	.927	72	.951	.309
24	.407	.914	74	.961	.276
26	.438	.899	76	.970	.242
28	.470	.883	78	.978	.208
30	.500	.866	80	.985	.174
32	.530	.848	82	.990	.139
34	.559	.829	84	.994	.104
36	.588	.809	86	.998	.070
38	.616	.788	88	.999	.035
40	.643	.766	90	1.00	.000
42	.669	.743			
44	.695	.719			
46	.719	.695			
48	.743	.669			

THREE-PLACE SINE AND COSINE TABLES

A	sinA	cosA	A	sinA	cosA
90	1.00	.00	140	.643	−.766
92	.999	−.035	142	.616	−.788
94	.998	−.070	144	.588	−.809
96	.994	−.104	146	.559	−.829
98	.990	−.139	148	.530	−.848
100	.985	−.174	150	.500	−.866
102	.978	−.208	152	.470	−.883
104	.970	−.242	154	.438	−.899
106	.961	−.276	156	.407	−.914
108	.951	−.309	158	.375	−.927
110	.940	−.342	160	.342	−.940
112	.927	−.375	162	.309	−.951
114	.914	−.407	164	.276	−.961
116	.889	−.438	166	.242	−.970
118	.883	−.470	168	.208	−.978
120	.866	−.500	170	.174	−.985
122	.848	−.530	172	.139	−.990
124	.929	−.559	174	.104	−.994
126	.809	−.588	176	.070	−.998
128	.788	−.616	178	.035	−.999
130	.766	−.643	180	.000	−1.00
132	.743	−.669			
134	.719	−.695			
136	.695	−.719			
138	.669	−.743			

THREE-PLACE SINE AND COSINE TABLES

A	sinA	cosA
180	.000	−1.00
182	−.035	−.999
184	−.070	−.998
186	−.104	−.994
188	−.139	−.990
190	−.174	−.985
192	−.208	−.978
194	−.242	−.970
196	−.276	−.961
198	−.309	−.951
200	−.342	−.940
202	−.375	−.927
204	−.407	−.914
206	−.438	−.899
208	−.470	−.883
210	−.500	−.866
212	−.530	−.848
214	−.559	−.829
216	−.588	−.809
218	−.616	−.788
220	−.643	−.766
222	−.669	−.743
224	−.695	−.719
226	−.719	−.695
228	−.743	−.669
230	−.766	−.643
232	−.788	−.616
234	−.809	−.588
236	−.829	−.559
238	−.848	−.530
240	−.866	−.500
242	−.883	−.470
244	−.899	−.438
246	−.914	−.407
248	−.927	−.375
250	−.940	−.342
252	−.951	−.309
254	−.961	−.276
256	−.970	−.242
258	−.978	−.208
260	−.985	−.174
262	−.990	−.139
264	−.994	−.104
266	−.998	−.070
268	−.999	−.035
270	−1.00	.00

THREE-PLACE SINE AND COSINE TABLES

A	sinA	cosA	A	sinA	cosA
270	−1.00	.000	320	−.643	.766
272	−.999	.035	322	−.616	.788
274	−.998	.070	324	−.588	.809
276	−.994	.104	326	−.559	.829
278	−.990	.139	328	−.530	.848
280	−.985	.174	330	−.500	.866
282	−.978	.208	332	−.470	.883
284	−.970	.242	334	−.438	.899
286	−.961	.276	336	−.407	.914
288	−.951	.309	338	−.375	.927
290	−.940	.342	340	−.342	.940
292	−.927	.375	342	−.309	.951
294	−.914	.407	344	−.276	.961
296	−.889	.438	346	−.242	.970
298	−.883	.470	348	−.208	.978
300	−.866	.500	350	−.174	.985
302	−.848	.530	352	−.139	.990
304	−.929	.559	354	−.104	.994
306	−.809	.588	356	−.070	.998
308	−.788	.616	358	−.035	.999
310	−.766	.643	360	.000	1.00
312	−.743	.669			
314	−.719	.695			
316	−.695	.719			
318	−.669	.743			

Appendix E

List of Constellations

Name	Abbreviation	Approximate Coordinates R.A.	dec.
Andromeda	AND	1^h	+35°
Antlia	ANT	10	−30
Apus	APS	3	−85
Aquarius	AQR	22	−05
Aquila	AQL	20	0
Ara	ARA	17	−55
Aries	ARI	2	+20
Auriga	AUR	5	+40
Bootes	BOO	15	+30
Caelum	CAE	4	−70
Camelopardalis	CAM	5	+70
Cancer	CNC	9	+20
Canes Venatici	CVN	12	+70
Canis Major	CMA	7	−20
Canis Minor	CMI	7	+05
Capricornus	CAP	21	−20
Carina	CAR	9	−60
Cassiopeia	CAS	1	+60
Centaurus	CEN	14	−60
Cepheus	CEP	22	+65
Cetus	CET	1	−12
Chamaeleon	CHA	9	−80
Circinus	CIR	15	−70
Columba	COL	6	−35
Coma Berenices	COM	13	+60

Name	Abbreviation	Approximate Coordinates	
		R.A.	dec.
Corona Australis	CRA	16^h	$-80°$
Corona Borealis	CRB	16	+30
Corvus	CRV	12	−20
Crater	CRT	11	−20
Crux	CRU	12	−60
Cygnus	CYG	20	+40
Delphinus	DEL	21	+15
Dorado	DOR	4	−55
Draco	DRA	16	+60
Equuleus	EQU	21	+10
Eridanus	ERI	4	−15
Fornax	FOR	3	−50
Gemini	GEM	7	+25
Grus	GRU	22	−40
Hercules	HER	17	+30
Horologium	HOR	3	−60
Hydra	HYA	11	−25
Hydrus	HYI	3	−75
Indus	IND	21	−60
Lacerta	LAC	22	+50
Leo	LEO	11	+20
Leo Minor	LMI	10	+40
Lepus	LEP	5	−20
Libra	LIB	15	−15
Lupus	LUP	15	−50

Name	Abbreviation	Approximate Coordinates R.A.	dec.
Lynx	LYN	8^h	+50
Mensa	MEN	6	−85
Microscopium	MIC	21	−40
Monoceros	MON	7	−05
Musca	MUS	13	−70
Norma	NOR	16	−65
Octans	OCT	21	−85
Ophiuchus	OPH	17	0
Orion	ORI	6	0
Pavo	PAV	21	−60
Pegasus	PEG	0	+20
Perseus	PER	3	+50
Phoenix	PHE	0	−45
Pictor	PIC	6	−50
Pisces	PSC	1	+10
Piscis Australis	PSA	23	−30
Puppis	PUP	8	−40
Pyxis	PYX	9	−40
Reticulum	RET	4	−60
Sagitta	SGE	20	+20
Sagittarius	SGR	19	−30
Scorpius	SCO	17	−30
Sculptor	SCL	0	−30
Scutum	SCT	19	−10
Serpens	SER	17	−

Name	Abbreviation	Approximate Coordinates R.A.	dec.
Sextans	SEX	10^h	$-05°$
Taurus	TAU	5	+20
Telescopium	TEL	18	−50
Triangulum	TRI	2	+30
Triangulum Australe	TRA	16	−65
Tucana	TUC	22	−60
Ursa Major	UMA	12	+55
Ursa Minor	UMI	16	+60
Vela	VEL	9	−50
Virgo	VIR	13	0
Volans	VOL	8	−80
Vulpecula	VUL	20	+25

Appendix F

Conversion Tables

CONVERSION TABLES

Minutes of Arc to Decimal Degrees
and
Minutes of Time to Decimal Hours

Minutes of Arc or Time	Decimal Degrees or Hours	Minutes of Arc or Time	Decimal Degrees or Hours
00	00	30	.50
02	.03	32	.53
04	.07	34	.57
06	.10	36	.60
08	.13	38	.63
10	.17	40	.67
12	.20	42	.70
14	.23	44	.73
16	.27	46	.77
18	.30	48	.80
20	.33	50	.83
22	.37	52	.87
24	.40	54	.90
26	.43	56	.93
28	.47	58	.97
30	50		

CONVERSION TABLES

Decimal Degrees to Minutes of Arc and Decimal Hours to Minutes of Time

Decimal Degrees or Hours	Minutes of Arc or Time	Decimal Degrees or Hours	Minutes of Arc or Time
.00	00	.50	30
.02	1	.52	31
.04	2	.54	32
.06	4	.56	34
.08	5	.58	35
.10	6	.60	36
.12	7	.62	37
.14	8	.64	38
.16	10	.66	40
.18	11	.68	41
.20	12	.70	42
.22	13	.72	43
.24	14	.74	44
.26	16	.76	46
.28	17	.78	47
.30	18	.80	48
.32	19	.82	49
.34	20	.84	50
.36	22	.86	52
.38	23	.88	53
.40	24	.90	54
.42	25	.92	55
.44	26	.94	56
.46	28	.96	58
.48	29	.98	59

Appendix G

Three-Place Antilogarithm Table

THREE-PLACE ANTILOGARITHM TABLE

Number	Alog	Number	Alog
0.1	1.26	1.6	39.8
0.2	1.58	1.7	50.1
0.3	2.00	1.8	63.1
0.4	2.51	1.9	79.4
0.5	3.16	2.0	100
0.6	3.98	2.1	126
0.7	5.01	2.2	158
0.8	6.31	2.3	200
0.9	7.94	2.4	251
1.0	10.0	2.5	316
1.1	12.6	2.6	398
1.2	15.8	2.7	501
1.3	20.0	2.8	631
1.4	25.1	2.9	794
1.5	31.6	3.0	1000

INDEX